AF618438

Modelling Trends in Waste Management

Debashish Sengupta · Sudha Agrahari
Editors

Modelling Trends in Solid and Hazardous Waste Management

Editors
Debashish Sengupta
Geology and Geophysics Department
Indian Institute of Technology Kharagpur
Kharagpur, West Bengal
India

Sudha Agrahari
Geology and Geophysics Department
Indian Institute of Technology Kharagpur
Kharagpur, West Bengal
India

ISBN 978-981-10-2409-2 ISBN 978-981-10-2410-8 (eBook)
DOI 10.1007/978-981-10-2410-8

Library of Congress Control Number: 2016958491

Printed on acid-free paper

This Springer imprint is published by Springer Nature
The registered company is Springer Nature Singapore Pte Ltd.
The registered company address is: 152 Beach Road, #22-06/08 Gateway East, Singapore 189721, Singapore

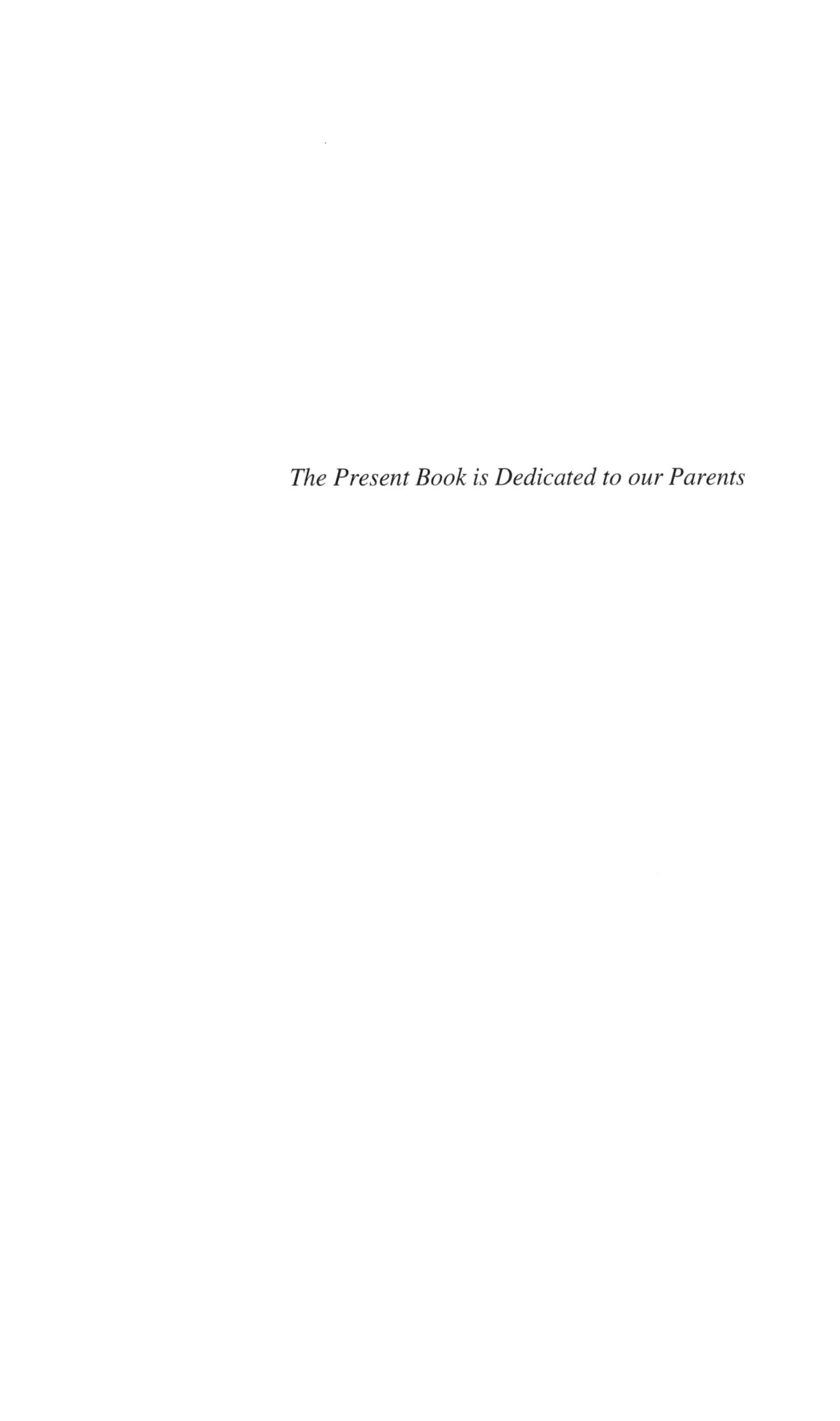

The Present Book is Dedicated to our Parents

Preface

The present book deals with the state-of-the-art research studies and technological developments in modeling trends and remediation of solid and hazardous waste management of various categories. **Hazardous waste** is a waste that creates substantial or potential threats to general health of public. Hazardous wastes are defined under RCRA in 40 CFR 261, according to which they have been divided into two major categories as characteristic hazardous and listed hazardous wastes. Characteristic hazardous wastes are materials that are tested for exhibiting four hazardous traits such as ignitability, reactivity, corrosivity, and toxicity. The hazardous wastes are materials listed by the regulatory authorities and connoted as hazardous wastes. These are from nonspecific sources, specific sources, or discarded chemical wastes. The various types of case studies discussed in this book include hospital waste which is an important type of urban waste. The waste from utilization and exploitation of energy resources enhances the heavy metal content and the ambient radiation environment 'Naturally Occurring Radioactive Materials (NORM)'. Procedural aspects for the framing of regulatory mechanism as employed for waste disposal and modeling of solid waste which includes municipal solid waste and the effects of hydrofracking for shale gas recovery is presented. The Non Invasive methods can be used for characterization and monitoring of solid waste and sustainable mineral development, primarily uranium resources for nuclear energy. The book also emphasizes the efficacy of Remote Sensing and GIS for geohazard studies, integrated geophysical methods for hazardous waste management and recent developments in site selection for landfills and innovations in microbial fuels cells.

The adequate protection and restoration of soil ecosystems contaminated by heavy metals require their characterization and remediation. Scattered literature is harnessed to critically review the possible sources, chemistry, potential biohazards, and best available remedial strategies for a number of heavy metals commonly found in contaminated soils. This book would be useful not only for practitioners of Environmental Science and Engineering but also to the readers who are involved in Environmental Planning and Management, preservation of heritage sites, development of suitable healthcare systems, and the practitioners in the framework of environmental-friendly "Green Technology." The context is of great significance due to the rapid urbanization and the development of smart cities for the growing population in various countries, especially developing countries like India, China, Brazil, South Africa, and Nigeria, to name a few.

In order to visually understand the trends in solid and hazardous waste management, we need the tools of remote sensing and GIS. GIS or geological information system is "location Intelligence" as propounded by the father of GIS Roger Tomlinson (1968). Necessity is the mother of invention. There is a dire need to understand the trends in solid and hazardous waste management for regional planning. This book encompasses the extensive research work by various research groups all over the world to understand the modeling trends in solid and hazardous waste management.

Kharagpur, India

Debashish Sengupta
Sudha Agrahari

Contents

About the Editors

Dr. Debashish Sengupta is Professor, Higher Administrative Grade and Ex-Head of the Department of Geology and Geophysics at the Indian Institute of Technology Kharagpur, West Bengal, India. Dr. Sengupta completed his Ph.D. in Applied Geophysics in 1987, and has more than 25 years of teaching and research experience. His areas of interests include nuclear geophysics with petroleum logging using subsurface nuclear data, radioactive methods and geochronology, radon emanometry and its applications, applications of isotopes and radionuclides in Earth and Environmental Geosciences, heat flow and geothermics. Dr. Sengupta has authored more than 90 research publications in international journals and more than 50 papers in conference proceedings, in addition to a large number of invited talks delivered both in India and abroad. His research work has been seminal and resulted in the formulation of environmental regulation policies in various countries of Asia, North and South America and the European Union. Dr. Sengupta had also been Visiting Professor at the University of Sao Paulo, Brazil and Senior Visiting Professor at the University of Salamanca, Spain. He received the Society of Geoscientists and Allied Technologists (SGAT's) Award of Excellence in Earth Sciences for the year 2003. To date, 10 students have been awarded Ph.D. degrees under his supervision.

Dr. Sudha Agrahari is Assistant Professor in the Department of Geology and Geophysics at the Indian Institute of Technology Kharagpur, India. She completed her Ph.D. in 2010, followed by three years of postdoctoral research experience in the field of helicopter-borne electromagnetics at the University of Cologne, Germany. Her research interests include unconventional energy resources, study of joint inversion of electrical and electromagnetic methods for environmental applications, and modeling and simulation of geophysical data. She has disseminated her research through publications in international journals and book chapters.

Management of Radioactive Wastes in a Hospital Environment

Ramamoorthy Ravichandran

Abstract Cytotoxic and radioactive wastes from hospitals are of major concern in the various types of urban wastes, and in developed countries they are managed by proper regulated methods. Nuclear medicine services are in the increase in many developing countries. Therefore to implement existing regulatory policies and guidelines in terms of handling of radioactive materials used in the treatment of patients and management of wastes arising out of hospitals need a good model. The commonly used radioisotopes in hospitals are: for diagnostic investigations ^{99m}Technetium (^{99m}Tc), ^{99}Molybdenum (parent for ^{99m}Tc), ^{67}Gallium, ^{201}Thallium, ^{131}I-MIBG, and ^{131}I (Capsules of activity 555MBq, 3.7,7.4 GBq(15,100,200mCi)) for therapeutic applications. Positron Emission radiopharmaceuticals do not pose much problem because of their short half lives. National regulatory authorities monitor the issues relating to organized safe use of radiopharmaceuticals in medical applications and give clearance for their use in hospitals. The sequences of protocols insisted are (1) Maintain inventory of receipts and safe custody of received radioisotopes, (2) Constant surveillance in terms of their safe applications during routine use (3) Safe methods of disposal of the radioactive wastes generated from human use of these radioisotopes and chemical formulations. Permissible concentrations of the radioactive waste disposals must take into account the community safety, and the expected degree of dilution is achieved at the discharge point of the institution into the sewage system. The hazard to the general population is insignificant in case the sludge containing radioactive waste material being used as fertilizer material. Depending on the type of radioactive waste from nuclear medicine procedures indicate to simply store the wastes till radioactive decay reduces the activity to a safe level or possibly by disposal of low-activity waste into the sewage system to achieve sufficient dilution. Disposal with permission from the regulatory authority and appropriate monitoring is known as a controlled disposal. Solid wastes (*Diagnostic*): Syringes, sharps, gloves kept in yellow plastic containers, allowed minimum 2 months decay, monitored by GM survey instrument and released. They bear dates written on them. They go for incineration. General

R. Ravichandran (✉)
National Oncology Center, Royal Hospital, Muscat, Oman
e-mail: ravichandranrama@rediffmail.com

D. Sengupta and S. Agrahari (eds.), *Modelling Trends in Solid and Hazardous Waste Management*, DOI 10.1007/978-981-10-2410-8_1

department wastes are sent only after 24 h elapsed, because most of them contain ^{99m}Tc cotton swabs etc., with half-life 6 h. Liquid wastes (*Diagnostic*): No laboratory work with open isotopes except for some labeling studies in the department. Diagnostic waiting patients' toilet urine wastes go to delay tank for increase in volume for I-131 patient's urine storage. Total about 2–4 months decay is ensured. ^{131}I *Therapy Isotope* is received in capsule form, and directly administered to patients orally. No liquid operations are carried out these days in the departments. I-131 (therapeutic) patients' linen, collected in an adjacent room in therapy ward, allowed to decay for varying periods 2–4 months; released to laundry after monitoring the count-rate, by an end-window type GM monitor. Half-life of ^{131}I is 8.05d, and therefore about 8–10 half lives will reduce the waste activity burden. ^{131}I therapeutic *Solid wastes* (nonactive black bags) (active yellow bags) all are monitored, and sent to '*Decay Waste Box*' kept in the backyard of the hospital, which is innovation by the author based on long experience in the hospital. Whenever the box becomes filled up (about 2–3 months), individual bags are re-monitored to ascertain residual radioactivity. If they reach background count rates, they are released. Waste release with BIN number, informed to regulatory authority for records. A special waste storage trolley help to keep the I-131 solid wastes from wards to allow decay, before these are released to waste treatment plant of the city. Some countries follow, centralized waste collection system which are commercial firms dealing with this operation throughout city, for many busy hospitals. Our local set up has a pair of delay tanks with individual volume total volume 36 M^3. The urine toilet releases are collected for about 2 months, kept closed for achieving a delay time of 2 months before releasing into the sewage treatment plant (STP) of the hospital. Six releases at 2 monthly intervals take place annually. At the time of releasing the contents of the delay tank, 50 mL samples are collected, and these sample are counted to document released activity of I-131 to confirm whether they are much below recommended levels of bi-monthly release. The released activity calculated showed good statistical correlation with measured values. The safe procedures followed for radioactive waste disposal is enumerated.

Keywords Radioactivity · Wastes management · Toxic wastes · Environmental pollution

Introduction

Waste management has become a global issue, and hospital wastes become subject of concern because of the associated risks due to their specific nature. The usual wastes released from the hospital are (a) solid and liquid infectious wastes, (b) sharps, (c) cytotoxic wastes, (d) pharmaceuticals, (e) chemicals and disinfectants, (f) pressurized containers, and (g) radioactive wastes. The applications of radionuclides in nuclear medicine for medical diagnosis and treatment have been steadily increasing ever since the discovery of artificial radioactivity. Radioactive wastes generate with the use of radioactive isotopes in large quantities, which has been either discarded because it is not required or not usable anymore. Handling unsealed sources leave some left over material from routine preparations, during dispensing of dose to patients, contaminated items resulting from handling them. The wastes in a large variety of forms based on the on the physical, chemical, and biological characteristics of the material. In medical and research work, there will be excreta or tissue specimens containing traces of radioactivity. The waste management aims at minimizing the risks in an effective way over the whole life cycle of the considered productions. Cytotoxic and radioactive wastes from hospitals are of major concern in the various types of urban wastes, and in developed countries they are managed by proper regulated methods. Nuclear medicine services are in the increase in many developing countries. Therefore to implement existing regulatory policies and guidelines in terms of handling of radioactive materials used in the treatment of patients and management of wastes arising out of hospitals need a good model. International Atomic Energy Agency (IAEA) has published manuals on the handling, transportation, and storage of radioactive wastes from application of radionuclides in hospitals. As most of the medical waste has a relatively short half-life, storage for decay on site is the key element in the optimization of waste management. This necessitates a well-organized selective collection, segregation, and packaging, following strict procedures and qualitative and quantitative measurement before and after decay. The clearance of nuclear waste after storage is considered a crucial step in the process. It should be based on careful control measurements. A brief account of methods practiced in a major hospital is presented.

Overview and Guidelines

It is customary to manage on site the biomedical radioactive wastes by decay storage. This reduce transport risks and ALARA (as low as reasonably achievable) activity levels. If the waste packages contain a mixed β, ν emitters the quantitative estimation of isotope activity becomes difficult. Therefore, segregation at the time of waste collection is recommended. Effective waste management indicates preparation of universal documentation procedures. These written procedures will deal with guidelines such as waste segregation at source in selected containers/ bags for waste accumulation. The staff members need to be trained well in the implementation of these guidelines. Department protocols indicate responsibilities at each level starting with management level with commitment for the waste policy. Data on all types of wastes generated, list of radionuclides, based on the procedures that are currently used for managing of radioactive wastes, use of decay storage and pretreatments of medical radioactive waste. Table 1 shows the radionuclides commonly used in clinical and biomedical research and found in wastes. Table 2 indicates short-lived Positron Emission radionuclides.

Table 1 Radionuclides in clinical and biomedical applications

Radionuclide	$T_{½}$	Emission and energy	Isotope	$T_{½}$	Emission and energy
H-3	12.3 Y	$E_{\beta max}$ 17 keV	Zn-65	244 d	β, ν 1.12 MeV
C-14	5730 Y	$E_{\beta max}$ 157 keV	Ga-67	3.3 d	β, ν 185 keV
P-32	14.3 d	$E_{\beta max}$ 1.17 MeV	Tc-99 m	6.0 h	β, ν 141 keV
S-35	87.5 d	$E_{\beta max}$ 167 keV	In-111	2.8 d	β, ν 171 keV
Cr-51	27.7 d	β, ν 320 keV	I-123	13 h	β, ν 159 keV
Co-57	272 d	β, ν 122 keV	I-125	60.1d	β, ν 27 keV
Co-58	70.8 d	β, ν 810 keV	I-131	8.05 d	β, ν 364 keV
Co-60	5.3 y	β, ν 1.25 MeV	Tl-201	3.4 d	β, ν 80 keV

Table 2 Positron emission radionuclides

Isotope	$T_{½}$	$\beta + E_{max}$ MeV	Annih. photons (keV)
C-11	20.4 min	0.959	511
N-13	10.0 min	1.197	511
O-15	2.0 min	1.738	511
F-18	110 min	0.633	511
Cu-64	12.7 h	0.650	511
Ga-68	68 min	1.898	511
Rb-82	75 s	3.400	511
Tc-94 m	52 min	0.755	511
I-124	4.2 d	2.130	511

The waste management strategies should be subject to periodic review, at least annually, to ensure optimization in waste management procedures, cost benefit analysis, and compliance with regulatory guidelines. If disposal routes change environmental impact, the individual disposal methods, i.e., incineration, landfill, low-level radioactive waste discharge to drainage, should be reviewed at the same time without violating compliance of regulatory guidelines keeping in mind the risks arising due to radiological and non-radiological impacts associated with type of wastes.

National and international recommendations dictate that radioactive waste management of the hospitals is one of the responsibility of the Radiation Safety Officer (RSO). Radioactive wastes arise in different forms depending on the variety of use to which the radionuclides are applied. Chemical operations generate wastes in solution form, precipitation, equipment decontamination, contaminated glass-ware, or agents applied for decontamination. Tissue specimens and excreta wastes results out of medical and biological work. The following types of radioactive wastes may occur in hospitals:

(i) Sealed sources
(ii) Excreta from patients treated with radionuclides
(iii) Unwanted solutions of radionuclides intended for therapeutic use
(iv) Normal low level liquid waste, e.g., from washing of apparatus or liquid scintillation counting residues
(v) Normal low level solid wastes, e.g., paper of glass
(vi) Wastes from spills and decontamination
(vii) Gases.

Wastes are classified into solid, liquid, and gaseous forms and finally the wastes are to be disposed into the urban sewage or atmosphere with prescribed limits by applying known methods. Basic guidelines toward waste disposal are available in a few hand books published from different countries. The permissible limits for different isotopes vary based on the physical, chemical, and radioactive toxicity considerations and their half lives. In India, Atomic Energy Regulatory Board (AERB) is the statutory/regulatory body of the Government of India, which gives policy issues from time to time relating to the radioactive waste disposal into the environment as well as public sewerage. At present the strategy are such that the user medical institution should prospectively plan in terms of the total radioactive wastes likely to be generated from the expected amount of radioactivity intended to be used, for the entire subsequent year. Authorization is necessary for the released wastes outside hospital premises. AERB under Radiation Protection Rules 1971, enforced by Atomic Energy Act, 1962. The document AERB Safety Code (SC-Med 4) [1] includes recommendations relating to disposal of radioactive wastes as below:

1. Stipulations for disposal of small quantities of radioisotopes as radioactive wastes by institutions
2. Guidelines for handling cadavers containing radionuclides administered before death.

An authorized person could carry out disposal of radioactive waste released into sanitary sewerage system if the wastes are in readily soluble form, and released quantity shall not exceed maximum limits shown in Table 3. The radioactive material released into the sewerage system by the institution per year shall not exceed 37 GBq. If many radionuclides are present the liquid waste, the sum of the fractions of the individual quantities of each of the radioisotopes shall not exceed respective maximum quantities allowed as in Table 3 added to unity. Details in a log book need to be maintained recording the isotopes disposed and the details of release.

Disposal of radioactive wastes is either by burial into pits or by incineration. Total activity buried per pit, the fractions of different isotopes buried should be based on the allowed limits in Table 4. The buried wastes should be below an earth layer of 1.20 m thickness, inter pit distance of 1.80 m, 12 burials per year, and reuse of pits

Table 3 Disposal limits for sanitary sewage system

Radionuclide	Maximum discharge limits MBq/day	Average monthly concentration in the discharge (MBq m^{-3})
H-3	92.5	3700
C-14	18.5	740
Na-24	3.7	222
P-32	3.7	18.5
S-35	18.5	74.0
Ca-45	3.7	10.1
Mo99 + Tc99 m	3.7	185
I-125	3.7	22.2
I-131	3.7	22.2

Table 4 Disposal limits for radioisotopes for ground burial

Radionuclide	Maximum activity in a pit (MBq)
H-3	9250
C-14	1850
Na-24	370
P-32	370
S-35	1850
Ca-45	370
Fe-59	370
Mo-99	370
I-125	37
I-131	37

only after completion of 10 half lives of the longest lived radioisotopes are some of the recommendations relating to burial. Special type of incinerators with provision for retention of solid and liquid combustion bye products arising out of radioactive wastes should be used. Air borne radioactive contaminations in excess of operation limits for unrestricted areas should not arise due to incineration. Release of solid and liquid wastes from incinerators should be collected and disposed off in accordance with the recommendations. Records of incineration and supervision by Radiological safety officer are additional requirements for radiation surveillance.

Diagnostic and Therapeutic Applications in Hospitals

The commonly used radioisotopes in a hospitals are: for diagnostic investigations ^{99m}Technetium (^{99m}Tc), ^{99}Molybdenum (parent for ^{99m}Tc), ^{67}Gallium, ^{201}Thallium, ^{131}Iodine-MIBG, and for therapeutic applications ^{131}I (Capsules of activity

555 MBq, 3.7 GBq, 7.4 GBq, (15, 100, 200 mCi)) for therapeutic applications. Positron Emission radiopharmaceuticals do not pose much problem because of their short half lives. National regulatory authorities monitor the issues relating to organized safe use of radiopharmaceuticals in medical applications and give clearance for their use in hospitals [11]. The sequences of protocols insisted are (1) Maintain inventory of receipts and safe custody of received radioisotopes, (2) Constant surveillance in terms of their safe applications during routine use (3) Safe methods of disposal of the radioactive wastes generated from human use of these radioisotopes and chemical formulations [2]. Permissible concentrations of the radioactive waste disposals must take into account the community safety, and the expected degree of dilution is achieved at the discharge point of the institution into the sewage system. The hazard to the general population is insignificant in case the sludge containing radioactive waste material being used as fertilizer material. Depending on the type of radioactive waste from nuclear medicine procedures indicate to simply store the wastes till radioactive decay reduces the activity to a safe level or possibly by disposal of low-activity waste into the sewage system to achieve sufficient dilution. Disposal with permission from the regulatory authority and appropriate monitoring is known as a controlled disposal. A position statement in a recently IAEA document indicated [4] that there is no need for storage of urine in delay tanks. Continuous sewage dilutions appears to be sufficient for I-131 urine release from therapy wards. However, in most of the places in the middle east Asian countries, there are no centralized sewage management systems connected to hospitals, and therefore, there is a need for controlled discharge of radioactive wastes, especially for radioactive liquid waste effluents.

Positron emission tomography (PET) radiopharmaceuticals do not pose much problem because of their short half lives. The radioactive waste generated from a PET department would typically consist of syringes, needles, i.v.lines, vials, gloves, cotton, tissue paper, absorbent papers, etc. The half-life of Fluorine-18 (F-18) is the longest, but 110 min (1 h 50 min), still considered short compared to conventional radioisotopes in the nuclear medicine department. This short half-life permits easy disposal of the radioactive waste associated with this isotope. The waste should be collected in shielded bins and after 10 half lives (about 20 h) decay, can be disposed of along with the general waste. Care should be taken to deface the radiation symbols on the vials and other articles, before they are discarded or recycled.

In medical cyclotron (MC) used for production of PET radioisotopes, various components of the cyclotron contain different elements which are subject to activation. A few of activated products are long lived to manifest radiation safety and waste disposal problem. Few of the activation products having long half life are V-49 (330 d), Mn-53 (3.74 M.y), Mn-54 (312.3 d), Co-57 (121.2 d), Ag-108 m (418 y), Ag-110 m (249.9 d), etc. A shielded and sealable storage area within the MC vault should be provided for housing these radioactive components [3]. Special handling procedures and care should undertaken while dealing with these activated components. The air inside the MC during irradiation can get activated. The presence of negative pressure inside the MC is essential to exhaust out this activity. With hospital-based MC, the presence of N-14, Ar-41 in the air exhausted into the atmosphere is believed to be of negligible hazard to the public. In the following sections, the mode of hospital waste management followed in Royal Hospital at Muscat, Sultanate of Oman is outlined. The applications of regulatory control on waste management could as well be appreciated.

Hospital Wastes Management

Diagnostic Wastes

A Royal Hospital Document [10] outlines procedures for infection control, also covering handling and disposal of radioactive wastes from diagnostic use of radioactive isotopes. Daily wastes arising out of ^{99m}Tc (disposables) are put along with normal wastes after 48 h delay. Plastic containers marked with wastes are used to collect used gloves and syringes. ^{67}Ga, ^{131}MIBG syringes are kept in different container and these long undergo minimum 2 months decay before these are released because they are relatively long-lived radioactive isotopes. They are released after monitoring by the department Medical Physicist using an end window GM survey meter. Department maintains the records of wastes released. A proforma will be faxed to medical wastes treatment plant (MWTP) when hospital releases any waste A copy of this document is also sent to the radiation protection adviser (RPA). To increase the volume of the I-131 delay tank (mentioned later), the toilet wastes of diagnostic patients go into delay tank, thereby achieving higher dilution factor.

I-131 Therapy Wastes

Royal Hospital imports I-131 capsules from GE Health Care, Amersham, UK. Presently no other radioisotope is used for therapy. An inventory of unused capsules is maintained in stores. After about 2 years elapsed they are allowed to go into the delay tank. Solid wastes from the isolation wards go into yellow and black bags. With marked patients' numbers they are sent to a temporary storage trolley (TST), which has provision for locking (Fig. 1). This is kept in the in the backyard of the hospital, and the contents are allowed to decay 2–3 months before disposal. After sufficient decay, the Individual bags will be checked with end window contamination probe. When the count rate is near background level, they are released for disposal, certified, and sent to MWTP, along with necessary documentation. As the volume of wastes increased, four boxes are being used, with sequential delay, decay, and disposal is followed.

Fig. 1 Temporary storage trolley at Royal Hospital, for decay

Delay Tanks

Two isolation rooms are available for administration of I-131 for therapy. Urine excreted from these patients contain about 90 % of administered radioactive iodine. These effluents got into separate sewage lines and reach the delay tanks. Figure 2 shows the two sewage connections from ward connected to twin concrete tanks located in the garden area below ground level. The size of the tanks are 5 m × 4 m × 2 m each, with 40 kL volume. Because of available ultrasound level sensors, only 75 % of volume could be filled viz. 30 kL when one tank is in operation. Two submersible pumps installed at 50 cm at the tank bottom make the flush volume of each delay tank as 24 kL.

Business management system (BMS) monitor levels of tanks, in a control system existing at the Engineering department, also showing status of opening, closing valves. A hooter also ensures maximum level and percentage filling. Certification procedure is followed for filling of delay tanks, closing and opening of valves, and releasing contents of the tanks. Filling phase of one tank will start when the other tank in filled condition to start decays for about 2 months [5–9]. Documents are maintained in the nuclear medicine and engineering departments regarding the status of valves and which tank (1 or 2) is in operation. We follow operational guideline as either 3.7 MBq/day or average monthly concentration 22.2 MBq/M^3 for maximum limit of discharge (refer Table 3). The BMS and manual sample collection before release of effluents are shown in Figs. 3 and 4.

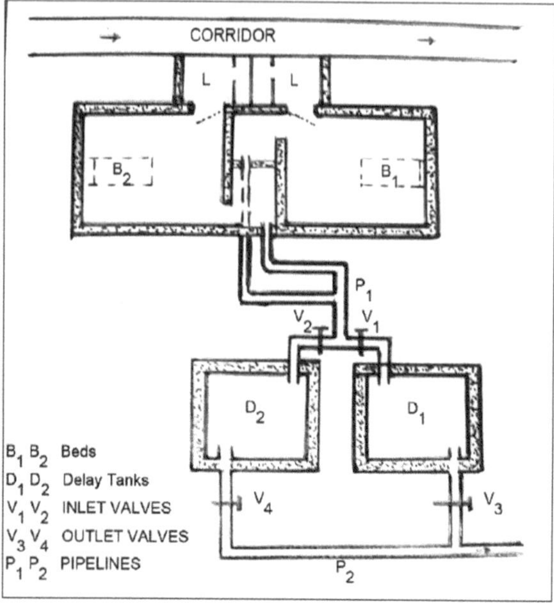

Fig. 2 Twin delay tanks at Royal Hospital for delay and decay

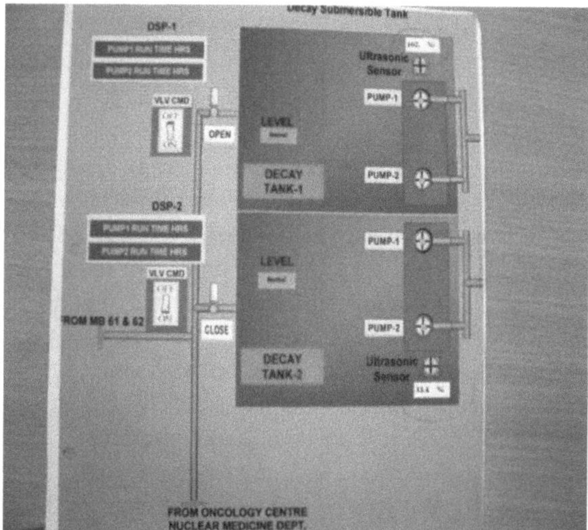

Fig. 3 BMS system display in monitor

Procedure for Release of Delay Tank Effluents

At the time of manual check, 50 mL effluent sample taken out from the delay tank is checked by end window contamination monitor. This is a gross check. One mL of this sample is separately counted by a well counter gamma ray spectrometer. Calibration of well counter is carried out by Cs-137 check source. Using counting

Fig. 4 Sample taken from delay tank

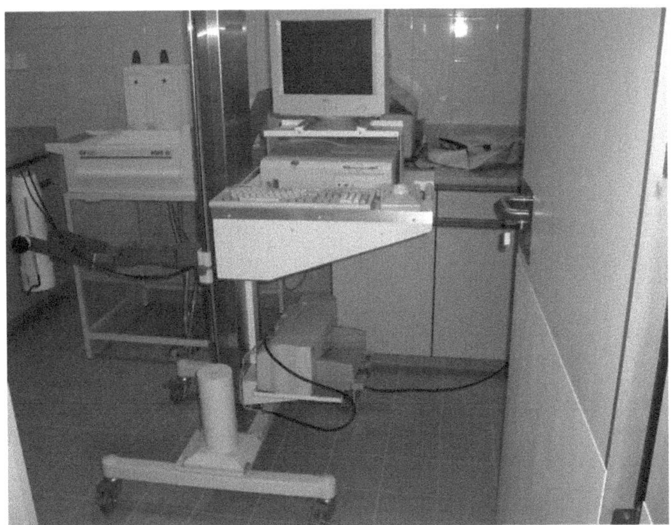

Fig. 5 Well counting system for estimation of I-131 activity/mL

efficiency of 30 % for this counting set up, the activity present in the delay tank sample is determined. Thc tap water one mL is also counted to express the true I-131 activity released in the total volume of 24 kL. Figure 5 shows the well

Table 5 Summary of patients admissions, filling, and delay phases relating to I^{131} procedures

Sl. No.	No. of patients	Date of closing	Total days		(MBq) at closing	Activity released (MBq)	
			Filling	Decay		Calculated	Measured
1	3	1.1.2006	123	55	20	0.2	0
2	4	3.5.2006	73	51	80	1	6.5
3	6	15.7.2006	63	78	3922	4.6	0
4	5	16.9.2006	78	51	315	3.8	0
5	7	3.12.2006	51	57	7223	79.8	84.5
6	7	24.1.2007	57	72	1774	3.5	2.8
7	8	24.3.2007	72	35	4038	194.7	145.8
8	9	4.6.2007	35	60	2654	14.7	1.5
9	7	10.7.2007	60	75	*Gonu*	2.8	12.1
10	9	10.9.2007	75	86	1846	1.6	6
11	7	25.11.2007	86	75	2762	5.2	0
12	10	19.2.2008	75	70	2660	10.6	4
13	8	4.5.2008	70	57	4544	7.5	7.4
14	8	14.7.2008	57	68	1047	5.1	7.7
15	9	8.9.2008	68	109	1834	0.4	0
16	8	16.11.2008	109	57	7161	54	55.9
17	8	1.3.2009	57	98	1530	0.5	0
18	8	8.6.2009	98	59	3850	8.5	0
19	6	7.10.2009	59	69	1418	7.2	4.8
20	5	23.1.2010	69	50	2826	37.6	118
21	10	6.4.2010	50	69	9869	11	17.3
22	5	20.6.2010	69	60	4680	41.11	44.8
23	12	14.8.2010	60	87	11483	5.9	12.3
24	8	11.11.2010	87	80	2970	3.7	0
25	7	31.1.2011	80	80	5619	75.9	118
26	12	22.3.2011	80	50	9587	50	42
27	10	22.5.2011	50	60	10896	3.7	5.3
28	12	22.8.2011	60	90	2963	1.9	8.1
29	10	29.11.2011	90	90	5972	3.7	0
30	8	4.3.2012	90	93	3073	7	506
31	7	13.5.2012	93	69	320	1.3	0
32	8	17.7.2012	69	64	2575	1	0
33	13	13.10.2012	64	86	8367	21.9	0
34	12	22.12.2012	86	69	8915	7.4	2.2
35	13	6.3.2013	69	74	9309	53.2	48
36	14	6.5.2013	74	60	9794	41	76
37	14	9.7.2013	60	63	7288	7.4	3.3
38	15	29.9.2013	63	80	8557	24	24
39	15	4.12.2013	80	65	7313	7.5	23.6
40	13	23.2.2014	65		10569	7.5	0

counting system for documenting the activity per mL of discharged delay tank decayed waste urine effluent. Table 5 gives the delay tank releases over period of 10 years demonstrating the efficacy of delay and decay method.

Discussion

Some countries follow, centralized waste collection system which are commercial firms dealing with this waste management throughout city, for many busy hospitals. From Table 5 it could be seen that external counting of samples showed good correlation to theoretical calculations based on total ward occupancy, activity administered and physical decay with decrement periods. Last two columns of Table 5 clearly brings out that never we exceeded operational limit of 22 MBq/m^3. Calculations for two admissions per session, four patients/week in two isolation wards, taking all patients administered with 5.5 GBq (150 mCi) has shown activity accumulation in the delay tank, 35.16 GBq. A further delay period of 2 months after closing valve in individual delay tank will envisage a decay factor of 0.5712 % (a factor of 0.005712) resulting in an activity to permissible levels. The temporary storage trollies for solid wastes of I-131, delay tank for liquid wastes, written policies, and documentation of waste disposal records are the solutions for the radioactive waste management in hospital environment.

Acknowledgements Thanks to Director General Royal Hospital; Head, Department of Nuclear Medicine; Principal Engineer, Royal Hospital for using data of releases of radioactive wastes.

References

AERB Safety Code No.SC/MED/4 for Nuclear Medicine Laboratories (1989) Revised in 2014

Applying radiation safety standards in nuclear medicine. Safety Report Series. No. 40. Vienna, IAEA; 2005

Eggermont GX, Covens P, Sonck M, Takacs S, Hermanne (2000). Optimization of nuclear hospital waste management. In: Rehani MM (ed) Advances in medical physics. Jaypee Brothers Med. Publication; New Delhi, pp 181–203

IAEA Position Statement (2010) Release of patients after radionuclide therapy. IAEA, Vienna

Ravichandran R, Jayasree U, Supe SS, Keshava SL, Devaru S (1997) Abstract, p 42. In: 23rd IARP conference recent advances in radiation measurements and radiation protection, Amristar, Feb 1997

Ravichandran R (1998) Storage and disposal of radioactive waste. In: Pant GS (ed) Radiation safety of unsealed sources, 1st edn. Himalaya Pub. Co., Mumbai, pp 102–114

Ravichandran R, Pant GS (2000) Storage and disposal of radioactive waste. In: Pant GS (ed) Radiation safety of unsealed sources, 2nd edn. Himalya Pub. Co., Mumbai, pp 237–248

Ravichandran R, ArunKumar LS, Sreeram R, Gorman K, Saadi AA (2006) Design, function, and radiation safety aspects of delay tank system connected to radioactive iodine isolation wads at oncology center. Oman J Med Phys 31:156–157

Ravichandran R, Binukumar JP, Sreeram R, ArunKumar LS (2011) An overview of radioactive waste disposal procedures of a nuclear medicine department. J Med Phys 36:95–99

Royal Hospital, (2001) Infection control. Policies and Procedures. The Infection Control Committee, pp 147–149

Thayalan K (ed) (2010) Radioactive waste disposal. In: Text book of radiological safety. Jaypee Med. Publication, Chennai, pp 267–288

Heavy Metal and Radionuclide Contaminant Migration in the Vicinity of Thermal Power Plants: Monitoring, Remediation, and Utilization

Debashish Sengupta and Sudha Agrahari

Abstract Environmental pollution has been an important issue particularly in a period which is witnessing a rapid industrial growth. The major sources of heavy metal and radionuclide pollution in urban areas are anthropogenic which includes the fossil, coal combustion, etc. Coal, one of the most abundantly found materials in nature, has trace quantities of primordial radionuclides, ^{238}U, ^{232}Th, and ^{40}K. Thus, combustion of coals in thermal power plants (TPPs) is one of the major sources of environmental pollution enhancing the natural radiation in the vicinity of TPPs. Fly ash, the main residue from the combustion of pulverized bituminous or sub-bituminous coal (lignite), is directly disposed of in large ponds in the nearby areas of TPPs. This problem is of particular significance in India, which utilizes coals of very high ash content (approximately 40–55 %). Background knowledge, i.e., monitoring of the origin, chemistry, and risks of toxic heavy metals (Pb, Cr, As, Zn, Cd, Cu, Hg, and Ni) and radionuclides in contaminated soils, is necessary for the selection of appropriate remedial options. For this purpose, soil, ash, and coal samples were collected from the nearly areas of TPPs. In order to validate the fact of heavy metal and radionuclide contaminant migration in the vicinity of TPPs, several geophysical, radiochemical, and trace element analysis tests were carried out. Lack of proper insulation at the bottom of the ash disposal ponds enhances the chances of groundwater contamination. The direct current (dc) resistivity survey, employing Schlumberger configuration, was undertaken to identify the local subsurface and to estimate the depth of contamination around ash ponds near a thermal power plant in Eastern India. A continuous conductive zone with resistivity $<5\ \Omega m$ was identified throughout the studied region at a depth of about 2–10 m indicating the presence of water with higher ionic concentration. The study was also supported by the radiometric geophysical measurements indicating the presence of radionuclides in the vicinity of ash ponds. Groundwater samples have also been collected

D. Sengupta (✉) · S. Agrahari
Department of Geology and Geophysics, Indian Institute of Technology Kharagpur, Kharagpur, India
e-mail: dsgg@gg.iitkgp.ernet.in; dsgiitkgp@gmail.com

S. Agrahari
e-mail: sudha@gg.iitkgp.ernet.in

D. Sengupta and S. Agrahari (eds.), *Modelling Trends in Solid and Hazardous Waste Management*, DOI 10.1007/978-981-10-2410-8_2

from the tube wells located near the ash ponds and analyzed for pH, TDS and trace elements. Results of the chemical analysis show high values of TDS and high concentration of the toxics. The studies on the geochemistry of the ash ponds have shown that the ashes are characterized by high concentration of As, Pb, Cu, Ni, Fe, Zn, Cr, Co, and Mn all of which exceeded the crustal abundance by a factor of 3–5. The high concentration of the toxics (As, Al, Li, Zn, Ag, Sb, Si, Mo, Ba, Rb, Se, Pb) in the water samples implies significant input from the ash pile due to leaching. This increases the TDS values of the water. Radiochemical characterization of coal, ash, and soil samples was done based on Radium equivalent activity, dose rate, and hazard index. The activity concentration variations infer to the high radioactivity in the vicinity of the ash ponds up to a radius of 500 m and indicate a decreasing trend beyond 500 m. In general, the radioactivity of the nearby area showed three to four times higher values than the prescribed limit, and the hazard index was estimated as unity. In order to mitigate the associated risks, make the land resource available for agricultural production, enhance food security, and scale down land tenure problems, remediation of contaminated soil is necessary. Immobilization, soil washing, and phytoremediation are one of the best technologies for cleaning up of contaminated soils but have been mostly demonstrated in developed countries. Such technologies need high recommendation for field applicability and commercialization in the countries (developing) where agriculture, urbanization, and industrialization are causing environmental degradation. The utilization of contaminated soil is also discussed. Rare earth elements, heavy metal, and radionuclides, which are of economical importance, can be extracted from the contaminated soil and the remainder can be utilized in filling voids and reclamation of the mining lands.

Keywords Thermal power plants · Radionuclides · VES · Radon exhalation

Introduction

One of the most significant fuel sources for steam and energy production is coal. Coal combustion in thermal power plants increases the three aspects of atmospheric pollution. Huge quantities of gaseous and particulate pollutants (SO_2, NO_x, CO_2, hydrocarbons, and fly ash) are discharged into the local environment. Additionally, small amounts of toxic trace elements (As, Hg, and Cd) and minute amounts of radionuclides (U, Th, and their radioactive daughters) are also released, which may have a radiological impact on the environment.

Coal is serving as a major power source in our country by fulfilling the 70 % electricity demand. Remaining part of power generation comes from the hydroelectricity and nuclear power plants. Coal ash is the residual after coal combustion, which was not fully burnt. The volume of the end residue generated after coal combustion is dependent on the quality of coal used in thermal power stations. In Indian thermal power stations, bituminous or sub-bituminous type of coal is used which is generating about 55–60 % ash content. Consequently, 160 million tons of ash is produced yearly and is projected to reach 300 million tons per annum by 2016–2017 (Singh et al. 2011). Because of the enrichment of radionuclides in coal ash, the management of coal ash is a major environmental problem.

Fly ash and bottom ash are mainly two types of solid wastes, which are generated as a byproduct after coal combustion. Trace elements, which are smaller in size and larger in surface area, got absorbed on the ash particles (Culec et al. 2001). Bottom ash is the coarse-grained fraction of ash, which precipitates due to its weight and settles at the bottom of the furnace. This course-grained fraction is directly thrown into the closer ash ponds. Due to the light weight of fly ash particles, it spreads and distributes in the vicinity by air, henceforth deposits on the soil surface, which increases the radioactivity level of the local environment.

The toxicity of the soil and groundwater gets enhanced due to continuous disposal of bottom ash into the ash ponds. The solid wastes generated after coal combustion are released either directly through the stack releases or indirectly from waste storage areas. Most of the power stations in India operate close to coal fields, which are mainly located in the eastern region of India, i.e., Bihar, Jharkhand, and West Bengal. Due to dense population density in these areas, the environmental aspect is of major concern. According to Lalit et al. (1986), the radioactivity level of coals from these coal fields is on the higher side compared to the other coal fields. Consequently, population residing in this particular region is facing more environmental hazard.

Out of 31 coal-based thermal power plants of eastern India, 14 of these are located in West Bengal. Kolaghat, Durgapur, and Bandel are one of the biggest thermal power plants of eastern India with a capacity of 1260, 350, and 530 MW of electricity, respectively. The Kolaghat thermal power plant is located in Midnapur district, West Bengal (22°27′–22°25′ N latitude and 87°50′–87°55′ E longitude). Coal used in this thermal power plant is of bituminous type and 1260 MW power is generated in six units of 210 MW. The Durgapur thermal power plant is situated in Burdwan district, West Bengal (22°56′–23°53′ N latitude and 86°48′–88°25′ E longitude), and Bandel thermal power plant in Hooghly district, West Bengal (23° 39′32′′–23°01′20′′ N latitude and 87°30′15′′–88°30′20′′ E longitude). Due to the use of low-grade coal (sub-bituminous), huge amount of ash is produced, which is thrown away in ash ponds. These ash ponds are open and exposed to the sun, and therefore the slurry starts drying and becomes hard and compact, which later becomes a source of radiation hazard.

Recently, the use of fly ash is increased many folds as building material in the form of ash bricks and as filling material for underground cavities. Thus it is evident to monitor the radiation risk to the population.

Here, we are going to discuss the three power plants (Bandel, Durgapur, and Kolaghat) in detail. For this purpose, these power plants were investigated (Mandal and Sengupta 2006; Maharana et al. 2010; Mondal et al. 2006; Parial et al. 2015, 2016). The general objectives were as follows:

i. The determination of the presence of radionuclides (U, Th, and K) in coal and ash, and their enrichments in ash after coal combustion. It gives an idea about the environmental hazard of the area.
ii. Because of the improper management of ash, it is directly dumped into the open ash ponds and thereby severely affecting the surrounding population. Therefore, the radiological impact is determined by computation of the dose rate.
iii. The determination of the subsurface contamination was done using noninvasive geophysical methods.
iv. A comparison between the radiometric data in the vicinity of Kolaghat, Durgapur, and Bandel thermal power plants is required in order to estimate the radiation potential in these areas.
v. Hazard index estimation for the fly ash admixture for cement, concrete, and fly ash bricks.

Materials and Methods

Characterization of the nearby areas of thermal power plants, in terms of heavy metals and radionuclide contaminant migration, is done by conducting field-based and lab-based analyses. Mostly field-based studies include noninvasive methods; however, lab-based analyses are done by conducting several geochemical tests. In this section, a brief summary of all the methods utilized is presented, which are used for the characterization of the vicinity of the TPPs.

1. ***Noninvasive Geophysical investigations:***

   i. ***In situ radiometric study***
   A reconnaissance survey was conducted in and around Kolaghat thermal power plant (KTPP) based on in situ radiometric analysis. The ambient terrestrial dose rate was measured using a handheld scintillation counter. The counter is commonly known as micro-R survey meter. The model used was UR-709, manufactured by Nucleonix Systems Private Limited, Hyderabad, India. Soil samples from high ambient dose rate areas were collected for lab-based radiometric analysis. The locations of the sampling points were marked using GPS (manufactured by Trimble, Juno ST Handheld, Trimble Juno Series).
   ii. ***Subsurface characterization using direct current (dc) resistivity method***
   1D vertical electrical sounding (VES) measurements were conducted using Terra Science resistivity meter. VES were measured at twelve locations around KTPP and at one reference site outside the contaminated area. These twelve soundings were conducted in the vicinity of the two ash ponds and within a radius of 1 km from the ash disposal sites using Schlumberger electrode configuration. To achieve greater depth of penetration, longest

possible profile at every location was selected. The measurements were carried out during winter and spring season to avoid the interferences due to rain. In dc resistivity sounding, current (I) is injected into ground through a pair of electrodes and the potential difference (ΔV) is measured by another pair of electrodes (potential electrodes). The placement of these two pairs of electrodes determines the array configuration. According to the Ohm's law, the ratio of the potential difference to the current injected estimates the resistance. The apparent resistivity is estimated using Ohm's law (Eq. 1):

$$\rho_a = G.\left(\frac{\Delta V}{I}\right) \tag{1}$$

where G is the geometric factor due to the array configuration, ΔV is the potential difference between the two potential electrodes, and I is the current injected. A detailed discussion can be found in Telford et al. (1990). The greater the current electrode separation, the greater is the depth of penetration.

2. ***Laboratory (Geochemical) investigations***

   i. ***Sample collection and preparation***

      The feed coal samples were collected from inside of the power plants. The coals from four different coal fields (Jharia, Talchir, Giridih, and Raniganj) of lower Gondwana age were mixed together and pulverized before combustion. The entire mass of generated ash in the power plants was disposed off in large ponds located near the power plant. Ash samples were collected from the ash ponds. Soil samples were collected from the villages in the vicinity of the active ash ponds. The collected samples were stored in polyethylene bags. The samples were homogenized and then dried for 24 h in an air circulation oven at 110 °C. About 100 g of the dried samples were weighed and sealed in gas-tight, radon impermeable, cylindrical polyethylene containers (7.0 cm height and 6.5 cm diameter). These samples were then left for 15 days in order to allow for radium, thorium, and their short-lived progenies to reach secular equilibrium before radioactive determination of uranium and thorium. Subsequently, gamma-ray spectrometric analyses of the samples were conducted for the estimation of the ^{238}U, ^{232}Th, and ^{40}K in soil, coal, and ash samples. The analysis was performed at the Radiochemistry Division, Variable Energy Cyclotron Centre, BARC, Kolkata.

   ii. ***Radioactivity measurements***

      Low-level gamma-ray spectrometry was used to determine the activity concentrations of ^{226}Ra, ^{232}Th, and ^{40}K in fly ash samples. The present system is an intrinsic p-type coaxial HPGe detector with a relative efficiency of 50 % and energy resolution of 1.9 keV at 1332.5 keV for ^{60}Co. It is placed vertically in a lead cylindrical shield of 12 cm thickness and 65 cm height to reduce the background radiation. A low ambient temperature is maintained by electric refrigeration system. A solid radionuclide mixture of gamma reference materials, sealed in standard cylindrical polyethylene containers (7.0 cm in height and 6.5 cm diameter), was used to estimate the absolute efficiency of the gamma-ray spectrometer.

a. *Estimation of Radiation dose and Hazard index*
   Natural radioactivity is a consequence of the radioactive emission of mainly ^{238}U, ^{232}Th, and ^{40}K. Potassium can be measured with the help of its own peak (1460 keV); however, uranium and thorium are measured with the help of their daughter products. For uranium the peaks 295 keV, 352 keV (^{214}Pb), and 609 keV (^{214}Bi), and for thorium the peaks of 338 keV, 911 keV (^{228}Ac), 239 keV (^{212}Pb), 727 keV (^{212}Bi), and 583 keV (^{208}Tl) are considered (Abd El-mageed et al. 2011; Khan et al. 2002). Efficiency of these respective peaks is calculated using a standard Eu^{152} sample, with known half-life, with the help of Eq. (2). Each sample is measured for 40,000 s to obtain a reasonable ϒ-ray peak area. From this peak area the activity is measured with the help of a multi-channel analyser Canberra DSA 1000, which is connected to computer through Genie 2 K software:

$$\varepsilon = \frac{CPS \times 100}{A_t \times I}, \tag{2}$$

   where ε = Efficiency; CPS = Counts per second; A_t = Activity of ^{152}Eu; and I = Intensity of ^{152}Eu.
b. *Absorbed Dose Rate*
   It is basically the radiation received by any person per kilogram of its mass. The conversion factor required to calculate absorbed dose rate at 1 m above the ground surface, due to uniform distribution of gamma radiations from ^{238}U, ^{232}Th, and ^{40}K, is calculated by the method proposed by the UNSCEAR 2000. The equation is mentioned below:

$$\mathrm{D\,(nGyh)} = 0.461\mathrm{A_U} + 0.623\mathrm{A_{Th}} + 0.0414\mathrm{A_K}, \tag{3}$$

   where $\mathrm{A_U}$, $\mathrm{A_{Th}}$, and $\mathrm{A_K}$ are the specific activities of ^{238}U, ^{232}Th, and ^{40}K in Bq kg^{-1}.
c. *Annual Effective Dose Rate*
   It basically gives us an idea about the internal and external radiation exposures. It is calculated using the conversion coefficients (0.7 SvGy^{-1}) and 0.2 as the outdoor occupancy factor, as proposed by UNSCEAR 2000:

$$\begin{aligned}\mathrm{AED}\left(\mathrm{mSvy^{-1}}\right) = \mathrm{D}\left(\mathrm{nGyh^{-1}}\right) \times 8760\left(\mathrm{hy^{-1}}\right) \\ \times 0.2 \times 0.7\mathrm{Sv\,Gy^{-1}} \times 10^{-6}.\end{aligned} \tag{4}$$

d. *Radium Equivalent*
It describes the gamma output from different mixtures of U, Th, and K in samples and helps in comparison of radiation exposure due to different radioisotopes:

$$Ra_{eq} = A_U + 1.43\ A_{Th} + 0.077\ A_K. \quad (5)$$

The equation is proposed by UNSCEAR 2000, based on the assumption that contributions from other radionuclides are insignificant.

e. *Hazard Index*
It is the index that indicates the external exposure due to the radioisotopes in the environment:

$$H_{ex} = A_U/370 + A_{Th}/259 + A_K/4810. \quad (6)$$

$H_{ex} < 1$ is considered to be within limit, and $H_{ex} > 1$ is unsafe and equivalent to 370 Bq kg^{-1}.

iii. ***Radon exhalation rate measurements***
Radon exhalation measurements were undertaken using Cup dosimeters. The fly ash samples were dried and sieved through a 100-mesh sieve and placed in cups/cans, up to the sensitive volume (diameter 7.0 cm and height 7.5 cm). An LR-115 type-II solid-state nuclear track detector (2 cm × 2 cm) was fixed at the top, within the cup. The sensitive, lower surface of the detector is freely exposed to the emergent radon and was capable of recording the alpha particles, due to the decay of radon in the cup. After a week or more, radon and its daughters attain equilibrium. Thus, the equilibrium activity of emergent radon could be obtained from the geometry of the cup and time of exposure. After exposure for 100 days, the detectors from all the cups were retrieved. For the revelation of tracks, the detectors were etched in 2.5 N NaOH at 60 °C for a period of 90 min, in a constant-temperature water bath. Alpha tracks on the exposed face of the detector foils were scanned using an optical microscope at magnification of 400 times. From the track density, the radon activity was obtained using a calibration factor of 0.056 Tr cm^{-2} d^{-1} obtained from an earlier calibration experiment, for an LR-115 type-II detector (Singh et al. 1997). An experiment was performed in order to study the effect on radon exhalation rates by adding varying amounts of fly ash. The measurement of the exhalation rate of radon from a homogeneous mixture of (i) fly ash and soil and (ii) fly ash and cement was obtained. For this purpose, different percentages of fly ash were introduced in the soil and cement, keeping the total weight of the sample constant. All the samples were dried and sieved through a 100-mesh sieve before mixing, for homogeneity.
The exhalation rate was estimated from the relationship provided by Fleischer and Morgo-campero (1978) and Khan et al. (1992):

$$E_x = \frac{CV\lambda}{A[T + 1/\lambda\{e^{-\lambda T} - 1\}]}, \quad (7)$$

where E_x is the radon exhalation rate (Bq m^{-2} h^{-1}), C is the integrated radon exposure measured by an LR-115 type-II plastic track detector (Bq m^{-3} h), V is the effective volume of Can (m^3), λ is the decay constant for radon (h^{-1}), T is the exposure time (h), and A is the area covered by the Can (m^2), respectively.

Results and Discussion

Environmental pollution measurements in and around TPPs were done by adopting two-step methodology. At first, the sites were investigated using noninvasive geophysical measurements and thereafter based on the preliminary noninvasive geophysical survey results, locations were selected for sample collection. As a second-step methodology, geochemical tests of soil/ash samples were conducted in laboratory.

i. ***Noninvasive Geophysical investigation results:***
   The isorad map, based on surface gamma radiation, from the vicinity of Kolaghat thermal power plant is shown in Fig. 1. The surface radiation varies from 4.5 to 9.2 μRhr^{-1} showing an increasing trend toward northwest. The maximum surface radiation dose was observed toward the northern edge of the

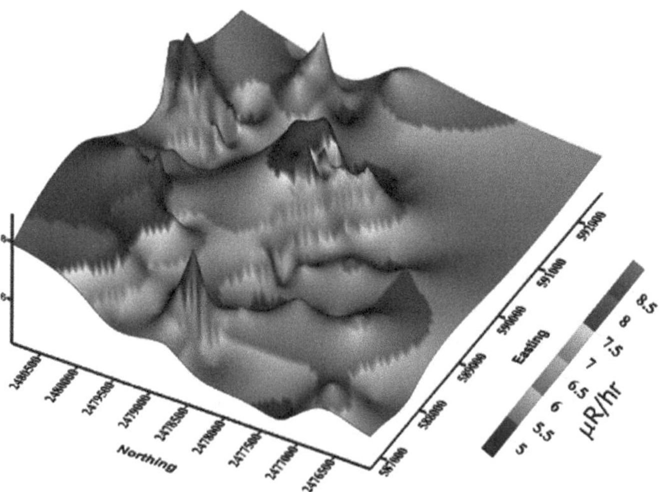

Fig. 1 Isorad map based on surface radiation in the vicinity of Kolaghat thermal power plant (after Parial 2016)

area, which coincided with the ash ponds present in the area. This indicated the presence of a high background radiation in the vicinity of the thermal power plant as well as the ash ponds (Parial 2016).

These observations emphasize that the contamination is maximum close to the ash ponds. The leachate from the ash ponds consists of water-soluble metallic species, which are posing maximum risk to groundwater. Generally, clay acts as a natural barrier to the leachates; however, due to high porosity sand not only acts as a potential aquifer but also allows an easy passage to the leachates from the waste dumps. Extreme precipitation events during monsoon can allow higher infiltration of the contaminants from the ash dumps to the deeper levels. Therefore, to assess the status of groundwater aquifers noninvasive vertical electrical sounding measurements were conducted in and around the Kolaghat thermal power plant. VES measurements using Terra Science resistivity meter (DDR2) were conducted at 12 locations and at one reference site. The reference site was located $\sim$3 km northwest of the thermal power plant; five soundings, VES-S4 to VES-S8, were in the vicinity of the two ash ponds; VES-S1 to VES-S3, VES-S9, VES-S10 were located within a radius of 1 km from the ash disposal sites. VES-S11 was located to the east of the thermal power plant, downstream of the River Rupnarayan, and VES-S12 was present south of the thermal power plant within 1 km (Fig. 2). The maximum profile length in the thirteen soundings varied between 100 and 800 m. The maximum profile length measured was at VES-S12 and the reference site.

These investigations help in understanding the subsurface in the region close to the thermal power plant. The 1D interpreted models infer the presence of a thick conductive zone (resistivity $< 5\ \Omega$m) throughout the region (Fig. 2). This conductive layer starts at a depth ranging from 2 to 10 m from the top and its thickness ranges between $\sim$27 and 39 m (Parial et al. 2015). Based on the available borehole data and the earlier observations (Mandal et al. 2007), this zone is identified as saturated clay with high concentration of various ionic species. From the borehole information, the presence of sandy layers at the top facilitates the percolation of the leachate into the deeper clay layer. Although clay is considered as a repository of trace/heavy metals, continuous loading of these species at higher concentration results in enhanced percolation. Ultimately, it percolates up to the deeper aquifer at a depth of $\sim$50 m (Fig. 2). As groundwater is the major source of potable water in this region, an extension with the previous studies by Mandal et al. (2007), Parial et al. (2015) emphasizes the importance of meticulous monitoring of the groundwater quality and the temporal variation in the nature of the contaminant(s).

ii. ***Geochemical investigation results:***

The geochemical studies of water samples, collected from a 50-m-deep tube well in the vicinity, by Mandal et al. (2007), further strengthened the presence of the mentioned contaminated layer ($<5\ \Omega$m). They confirmed the presence of high TDS count (550–760 ppm), abundance of Ca^{2+}, Na^{+}, presence of anions (HCO_3^-, SO_4^{2-}, and Cl^-), and high concentration of trace elements (like Li,

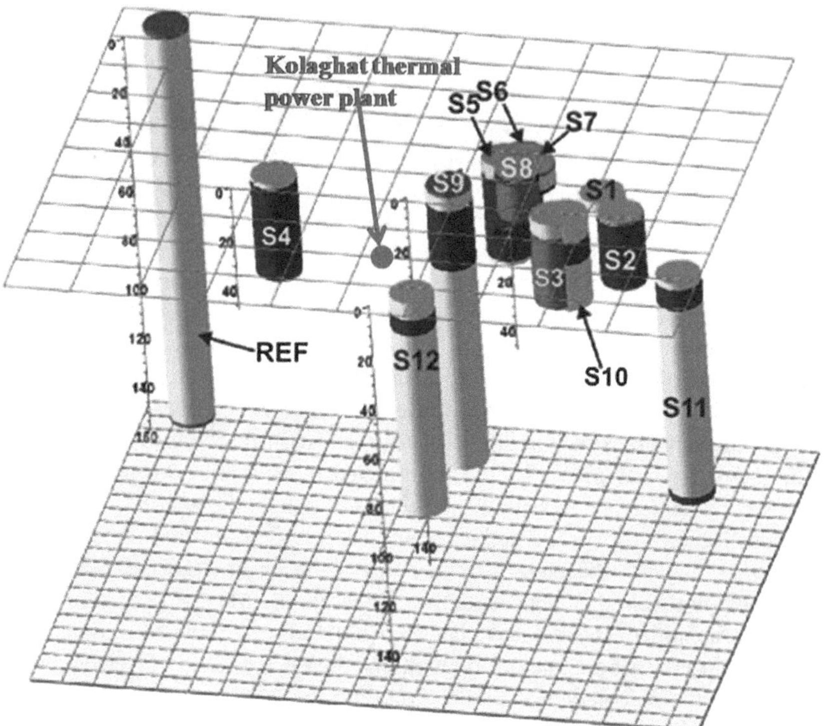

Fig. 2 Geoelectric section from the vicinity of Kolaghat thermal power plant derived from interpreted resistivity and borehole log (modified after Parial et al. 2015)

As, Zn, Ag, Pb, Cr, Mn, Se, Cd, Cu) indicating the high concentration of various ionic species within the clay layer.

We analyzed the ash samples from three power stations located in West Bengal (Kolaghat, Durgapur, and Bandel). Tables 1, 2, and 3 show the results of radiometric analysis from these power stations (Mondal et al. 2006). Kolaghat ash showed the activity concentration variation of ^{232}Th, ^{238}U, and ^{40}K in the range from 126 to 146 Bq kg^{-1} (average 140 Bq kg^{-1}), 98–119 Bq kg^{-1} (average 111 Bq kg^{-1}), and 266–415 Bq kg^{-1} (average 351 Bq kg^{-1}), respectively. Similarly, the activity concentration variations of ^{232}Th on

Table 1 Radioactivity measurements in fly ash samples from Kolaghat thermal power station, West Bengal (after Mondal et al. 2006)

Sample No.	^{238}U (Bq kg^{-1})	^{232}Th (Bq kg^{-1})	^{40}K (Bq kg^{-1})	Dose rate (nGy h^{-1})	Annual external effective dose rate (mSv y^{-1})	Ra_{eq} (Bq kg^{-1})	H_{ex} (Bq kg^{-1})
KAP1	117.4	145.8	403.0	159.1	0.195	357	0.96
KAP2	119.8	146.7	279.0	155.6	0.191	351	0.95
KAP3	114.9	144.2	282.1	151.9	0.186	343	0.92
KAP4	117.4	143.0	266.6	151.8	0.186	342	0.92
KAP5	98.9	141.8	372.0	146.8	0.180	330	0.89
KAP6	110.0	146.7	272.8	150.8	0.185	341	0.92
KAP7	117.5	136.2	372.0	152.1	0.187	341	0.92
KAP8	108.0	136.4	372.0	147.8	0.181	332	0.89
KAP9	113.7	126.1	359.6	143.7	0.176	322	0.87
KAP10	108.4	138.7	384.4	149.9	0.184	336	0.91
KAP11	107.2	139.7	368.9	149.3	0.183	335	0.90
KAP12	107.5	146.3	415.4	155.3	0.191	348	0.94
KAP13	113.9	141.4	403.0	154.9	0.190	347	0.94
KAP14	105.2	129.7	359.6	141.9	0.174	318	0.86
Mean	111.4	140.2	350.7	150.8	0.185	339	0.91

Durgapur and Bandel Thermal Power Stations indicate an average value of 107 and 106 Bq kg^{-1} (ranged from 62 to 139 Bq kg^{-1}), respectively.

On comparing the variation in radionuclides activity concentration measured at these power stations, it indicated that the enrichment of a particular radionuclide in the ashes is different at different power stations. Here, Kolaghat ashes indicate maximum enrichment of ^{232}Th compared to Durgapur and Bandel ashes. However, Bandel ashes are enriched in ^{238}U. Durgapur ashes show lesser activity concentration for ^{40}K compared to Kolaghat and Bandel ashes. Ashes from these power stations show more activity concentration for ^{232}Th compared to ^{238}U activity concentration. The feed coal analysis from these power plants also indicates the maximum activity concentration of ^{232}Th compared to ^{238}U (Mandal and Sengupta 2003). These power plants infer to the high radionuclides activity concentration compared to the power plants located in other parts of India. Table 4 shows the mean activity coefficients of radionuclides (U, Th, and K) and absorbed dose rates of fly ashes from other thermal power plants (Mondal et al. 2006).

Due to the presence of uranium in the organic material in sedimentary rocks, it accumulates in coal, during the formation of coal. Uranium is mainly retained in the carbonaceous part, due to sorptive uptake during the initial stages of peat accumulation and subsequent burial (Zielinski et al. 1986). In contrast, the inorganic phases retain most of the thorium. The geological formation around the ash ponds

Table 2 Radioactivity measurements in fly ash of Durgapur thermal power plant (after Mondal et al. 2006). The annual external effective dose rate (mSv y^{-1}), radium equivalent activity (Ra_{eq}), and external hazard index (H_{ex}) have also been provided

Sample No.	^{238}U (Bq kg^{-1})	^{232}Th (Bq kg^{-1})	^{40}K (Bq kg^{-1})	Dose rate (nGy h^{-1})	Annual external effective dose rate (mSv y^{-1})	Ra_{eq} (Bq kg^{-1})	H_{ex} (Bq kg^{-1})
DAP1	96.6	109.1	300.7	123.1	0.142	276	0.74
DAP2	93.1	105.3	280.1	118.3	0.142	265	0.71
DAP3	96.5	109.3	290.4	122.7	0.145	275	0.74
DAP4	94.1	101.0	267.3	115.7	0.159	259	0.70
DAP5	85.4	107.6	272.4	115.8	0.146	260	0.70
DAP6	96.5	98.1	354.7	118.6	0.157	264	0.71
DAP7	100.4	113.9	344.4	129.5	0.150	290	0.78
DAP8	95.6	103.8	287.8	118.8	0.157	266	0.72
DAP9	100.4	110.0	364.9	128.1	0.143	286	0.77
DAP10	99.6	104.0	326.4	122.4	0.158	273	0.74
DAP11	101.7	112.0	321.3	128.0	0.161	287	0.77
DAP12	99.3	96.0	313.5	116.9	0.156	261	0.70
DAP13	102.9	112.0	334.1	129.1	0.142	289	0.78
DAP14	96.7	120.0	336.7	131.2	0.142	294	0.79
DAP15	100.4	110.3	341.8	127.3	0.145	284	0.77
Mean	97.3	107.8	315.8	123.0	0.150	275	0.74

Table 3 Radioactivity measurements on fly ash samples of Bandel thermal power plant (after Mondal et al. 2006). The Ra equivalent activity (Ra_{eq}) and external Hazard index (H_{ex}) are also provided

Sample No.	^{238}U (Bq kg^{-1})	^{232}Th (Bq kg^{-1})	^{40}K (Bq kg^{-1})	Dose rate (nGy h^{-1})	Annual external effective dose rate (mSv y^{-1})	Ra_{eq} (Bq kg^{-1})	H_{ex} (Bq kg^{-1})
BAP1	116.7	139.9	327.7	152.0	0.186	342	0.92
BAP2	128.4	123.9	296.9	146.6	0.180	328	0.89
BAP3	133.4	111.6	332.8	142.9	0.175	319	0.86
BAP4	78.5	62.8	307.2	87.0	0.107	192	0.52
BAP5	183.4	118.2	174.1	163.4	0.200	366	0.99
BAP6	130.2	94.8	417.3	134.8	0.165	298	0.80
BAP7	118.3	92.6	396.8	127.1	0.156	281	0.76
Mean	126.9	106.3	321.8	136.3	0.167	304	0.82

Table 4 Mean activity of uranium, thorium, and potassium concentration along with absorbed dose rates of fly ash samples from some thermal power plants in India

Thermal power stations (India)	Activity (Bq kg^{-1})			Absorbed dose rates (nGt h^{-1})	Reference
	^{226}Ra	^{228}Ac	^{40}K		
Allahabad (Uttar Pradesh)	78.4	89.1	362.7	107.59	Vijayan and Behera (1999)
Angul (Odisha)	78.5	86.5	278.1	102.38	Vijayan and Behera (1999)
Badarpur (Delhi)	75.5	88.1	286.4	102.50	Vijayan and Behera (1999)
Chandrapur (Madhya Pradesh)	58.2	89.2	301.2	96.46	Vijayan and Behera (1999)
Raichur (Karnataka)	83.1	102.5	334.1	117.27	Vijayan and Behera (1999)
Talchir (Odisha)	79.2	96.3	291.6	109.73	Vijayan and Behera (1999)
Bokaro (Bihar)	70.3	118.4	252.0	118.91	Lalit et al. (1986)
Ramagundam (Andhra Pradesh)	59.2	95.1	507.0	109.38	Lalit et al. (1986)
Neyvelli (Tamil Nadu)	64	126.9	370.0	126.76	Lalit et al. (1986)
Amarkantak (Madhya Pradesh)	49.2	106.2	329.3	105.04	Lalit et al. (1986)
Nasik (Maharashtra)	126.9	138.0	279.0	157.18	Lalit et al. (1986)
Nellore (Andhra Pradesh)	64	126.9	370.0	126.76	Lalit et al. (1986)
Farakka (West Bengal)	84.1	98.8	297.1	113.71	Vijayan and Behera (1999)
Bakreshwar (West Bengal)	76.3	87.5	288.1	102.52	Vijayan and Behera (1999)
Kolaghat (West Bengal)	111.4	140.2	350.7	150.8	Mondal et al. (2006)
Durgapur (West Bengal)	97.3	107.5	315.8	123.0	Mondal et al. (2006)
Bandel (West Bengal)	126.9	106.3	321.8	136.3	Mondal et al. (2006)

of these power plants is mainly laterites overlained by clay. Clays are considered as potential trace elemental repositories. Therefore, the mixture of dry ashes and clay forms the top soil around these three thermal power plants. This facilitates the adsorption of radionuclides on the top soil (clayey layer). Due to the volatile nature of uranium, it migrates deeper by the percolating rain water and subsurface run-off; however, thorium remains adsorbed on the clayey formations. This is the reason that the ash samples show high activity concentration for ^{232}Th compared to ^{238}U. Hence, the dose rates of the measured ash samples are following the trend of thorium activity concentration, which mean high concentration thorium contributes more in the dose rate. The activity concentration of the natural radionuclides in feed coal samples is 2–3 times lesser than the ash samples (Fig. 3).

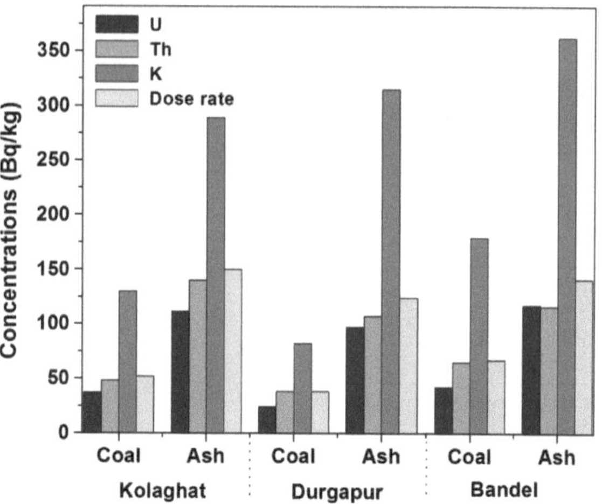

Fig. 3 Radioactivity enhancement from coal to ash in Kolaghat, Durgapur, and Bandel thermal power plants (after Mondal et al. 2006)

To estimate the external gamma dose and internal dose due to radon and its daughters, Ra_{eq} is calculated (Tables 1, 2, and 3) for the ash samples. The maximum permissible value for Ra_{eq} is 370 Bq kg^{-1} and for external hazard index (H_{ex}) is unity. These values are calculated using Eqs. (5) and (6).

In order to monitor the migration of contaminants laterally, several soil samples were collected and analyzed. High radioactivity was observed within a radius of 500 m and it decreases as we move further away from the ash ponds (Parial et al. 2016).

Nowadays, fly ash is being utilized in civil constructions in several forms; therefore, the radiation potential estimation becomes necessary (Eisenbud and Gesell 1997; MacKenzie 2000; UNSCEAR 1993, 2000). On comparison with coal, enhanced levels of uranium are present in fly ash (Jojo et al. 1993). Therefore, it is not easy to calculate the radon exhalation rate by uranium's or its daughter product's concentration due to its dependence on the texture and grain size composition (Tufail et al. 1991). Interestingly, the radon exhalation values are more in coal samples compared to fly ash samples. This may be due to the difference in the texture of coal and fly ash. In addition, the finer fly ash particles fill the small voids in the material and check the radon escape.

Easy availability and low cost of fly ash makes it an attractive substitute as building materials. In order to limit the radiation potential, it can be used as an additive in building materials. Several studies are reported which show the influence of fly ash as an additive on radiation level. However, different results are reported by different groups. Some authors reported a decreasing trend in radon exhalation in the mixture of concrete and fly ash additives (Stranden 1983; Maraziotis 1985). On the contrary, others have reported opposite trend when fly ash was used as an additive in cement (Siotis and Wrixon 1984). Some case studies reported no significant change in radon exhalation in cement and soil samples after mixing of fly ash in it (Ulbak et al. 1984; Karmadoost et al. 1988).

Table 5 Radon exhalation rate in cement and soil with different proportions of fly ash (expressed in weight percent) added in the samples of equal total weight (Kumar et al. 2005)

Sample	Fly ash additive (%)	Track density (cm^{-2} d^{-1})	Radon activity (Bq m^{-3})	Radon exhalation rate (mBq m^{-2} h^{-1})	Expected values from an arithmetic mean
1. Cement					
AC0	0	26.8 ± 1.5	482 ± 26	173 ± 9	–
AC1	10	49.9 ± 2.0	891 ± 36	320 ± 13	207
AC2	20	65.9 ± 2.3	1176 ± 41	422 ± 15	241
AC3	30	67.5 ± 2.3	1206 ± 42	433 ± 15	275
AC4	40	69.2 ± 2.4	1235 ± 42	443 ± 16	310
AC5	50	79.1 ± 2.5	1412 ± 45	507 ± 17	344
2. Soil					
AS0	0	100.9 ± 2.8	1803 ± 51	648 ± 18	–
AS1	10	91.3 ± 2.7	1630 ± 49	586 ± 18	634
AS2	20	75.5 ± 2.5	1348 ± 44	484 ± 16	621
AS3	30	72.24 ± 2.4	1290 ± 43	463 ± 15	608
AS4	40	65.16 ± 2.3	1163 ± 41	418 ± 15	595
AS5	50	60.3 ± 2.2	1076 ± 39	387 ± 14	582

To obtain a statistically reliable correlation on the application of fly ash as an additive, the '*t*-test' was used on the measurements obtained. The results for a homogeneous mixture of (i) cement and fly ash and (ii) soil and fly ash, in varying amounts, are summarized in Table 5. The samples with fly ash as additive in cement samples have shown a gradual increase, whereas a decrease is seen in soil samples, subsequent to the addition of fly ash. The experimental values for mixtures of ash and cement are higher than the values expected from the computations based on relative proportions of ash and cement, respectively. An opposite trend was observed for mixtures of ash and soil. This indicates that the values for the admixture increase, if the additive has higher exhalation rate than the base material and decrease if the value for additives is less (Kumar et al. 2005). This has been attributed to the emanating powers of different materials and also due to the difference in the grain size. The results obtained are significant, in terms of appropriate suppression of radon.

Fly Ash Remediation and Utilization

Both India and China have the highest fly ash production per year. However, the amount utilized is about 38–45 % only. This is compared to about 65 % fly ash utilization in U.S.A. (Basu et al. 2009; other references cited there-in).

The use of fly ash as agricultural amendment, fly ash bricks, cement, and concrete and in pavements and highways is of considerable significance. In addition, they are being used in production of glasses and glass-ceramics (Barbieri et al. 1999), as well as utilized to construct fills and embankments, grouts for pavements, structural fills, and retaining walls (Lamb 1974). More recently coal ashes have also been suggested as potential sources for rare earth extraction including yttrium (Kashiwakura et al. 2013; Seredin and Dai 2012).

The Organization for Economic Cooperation and Development (OECD) has classified the coal ash byproducts as Green List waste. Under Basel Convention it is not considered as a waste. The coal ash byproducts have not been utilized in many countries; however, they have been neglected like a waste material. Lack in the availability of appropriate cost-effective technologies prevents the optimum utilization of fly ash in India (Bhattacharjee and Kandpal 2002).

Geologically, various toxic elements and natural radionuclides are present in coal. Therefore, after burning the coal, the residues either gaseous emission or solid wastes (fly and bottom ash) or liquid discharge (leachates from ash ponds) are having these elements. According to Gabbard (1993), burning of the coal reduces the coal volume by 85 %, which mean that the radionuclides and trace elements are more concentrated in the residues (Technologically Enhanced Naturally Occurring Radioactive Material, TENORM). The partitioning behavior and fate are decided by combustion environment and the association of these elements with different phases/matrixes in coal (Papastefanou 2010; Karangelos et al. 2004). Due to the presence of ^{232}Th in inorganic matrix and ^{238}U in organic matrix, ^{238}U and ^{232}Th partition differently during combustion (Menon et al. 2011). ^{238}U shows a preference for fine to ultrafine ash particles and are mostly released in the ambient environment through the stack (Papastefanou 2010; Coles et al. 1979). However, ^{232}Th shows no preferential partitioning (Papastefanou 2010). The overall impact due to the radioactivity of the ash thus is mainly dependent on the type and amount of coal burnt for the energy generation. Therefore, the type and amount of coal burnt in the power production will be the deciding factor for radiation impact due to the radioactivity of the end products.

Conclusions

The radioactivity analysis presented in this chapter shows considerable amount of contamination of the surface soil in the vicinity of the ash ponds. It has also been observed that the surface soil contamination is more prominent in the predominant wind direction. The background radioactivity, calculated from the emitted dose, of the ambient environments of Kolaghat, Durgapur, and Bandel thermal power plants is approximately three times higher than the world average of 43 $nGyh^{-1}$. The ashes from these power plants have 2–3 times more concentration of natural radionuclides than feed coal. Consequently, the populations living in the nearby areas are at higher health risk. The radium equivalent activity (Ra_{eq}) and external hazards index (H_{ex}) values are close to 370 Bq kg^{-1} and unity, respectively. Therefore, the amount of ash used as additives in building materials should be examined cautiously in order to keep a check on the concentration of natural radionuclides.

Among the leading available technologies for the remediation of heavy metal contaminated soils are immobilization, soil washing, and phytoremediation but have been frequently practiced in developed countries. These technologies are suggested for field applicability and commercialization in the developing countries, where environmental degradation is the rancorous legacy of the urbanization, industrialization, and agricultural activities.

References

Abd El-mageed AI, El-Kamel AH, Abbady A, Harb S, Youssef AMM and Saleh II (2011) Assessment of natural and anthropogenic radioactivity levels in rocks and soils in the environments of Juban town in Yemen. Radiat Phys Chem 80:710–715

Barbieri L, Lancellotti I, Manfredini T, Queralt I, Rincon JM, Romero M (1999) Design, obtainment and properties of glasses and glass ceramics from coal fly ash. Fuel 78(2):271–276

Basu M, Pande M, Bhadoria PBS, Mahapatra SC (2009) Potential fly-ash utilization in agriculture: a global review. Prog Nat Sci 19:1173–1186

Bhattacharjee U, Kandpal TC (2002) Potential of flyash utilisation in India. Energy 27:151–166

Coles DG, Ragaini RC, Ondov JM, Fisher GL, Silberman D, Prentice BA (1979) Chemical studies of stack fly ash from a coalfired power plant. Environ Sci Technol

Culec N, Gunal (Calci) B, Erler A (2001) Assessment of soil and water contamination around an ash-disposal site: a case study from the Seyitomer coal-fired power plant in western Turkey. Environ Geol 40:331–344

Eisenbud M, Gesell T (1997) Environmental radioactivity, 4th edn. Academic Press, San Diego

Fleischer RL, Morgo-campero A (1978) Mapping and integrated radon emanation for detection of long distance migration of gases within the earth, techniques and principles. Geophys Res 83:3539–3549

Gabbard A (1993) Coal combustion—ORNL review. Oak Ridge Natl Lab Rev 26(3&4):24–33

Jojo PJ, Rawat A, Kumar A, Prasad R (1993) Trace uranium analysis in Indian coal samples. Nucl Geophys 7:445–448

Karangelos DJ, Petropoulos NP, Anagnostakis MJ, Hinis EP, Simopoulos SE (2004) Radiological characteristics and investigation of the radioactive equilibrium in the ashes produced in lignite-fired power plants. J Environ Radioact

Karmadoost NA, Durrani SA, Fremlin JH (1988) An investigation of radon exhalation from fly ash produced in the combustion of coal. Nucl Tracks Radiat Meas 15:647–650

Kashiwakura S, Kumagai Y, Kubo H, Wagatsuma K (2013) Dissolution of rare earth elements from coal fly ash particles in a dilute H2SO4 solvent, Open J Phys Chem 3:69–75

Khan K, Aslam M, Orfi SD, Khan HM (2002) Norm and associated radiation hazards in bricks fabricated in various localities of the North-West Frontier Province (Pakistan). J Environ Radioact 58:59–66

Khan AJ, Prasad R, Tyagi RK (1992) Measurement of radon exhalation rate from some building materials. Nucl Tracks Radiat Meas 20:609–610

Kumar R, Mahur AK, Sengupta D, Prasad R (2005) Radon activity and exhalation measurements in fly ash from a thermal power plant. Radiat Meas 40:638–641

Lalit BY, Ramachandran TV, Mishra UC (1986) Radiation exposures due to coal-fired power stations in India. Radiat Protect Dosimetry 15:197–202

Lamb DW (1974) Ash disposal in dams, mounds, structural fills and retaining walls. In: Proceedings of the third international ash utilization symposium. U.S. Bureau of Mines, Information Circular No. 8640, Washington DC

MacKenzie AB (2000) Environmental radioactivity: experience from the 20th century-trends and issues for the 21st century. Sci Total Environ 249:313–329

Maharana, M, Eappen KP, Sengupta D (2010) Radon emanometric technique for ^{226}Ra estimation. J Radioanal Nucl Chem 285(3):469–474

Mandal A, Sengupta D, Sharma SP (2007) DC Resistivity studies for mapping groundwater contamination in and around ash-disposal site of Kolaghat thermal power plant, West Bengal. J Geol Soc India 69:373–380

Mandal A, Sengupta D (2006) An assessment of soil contamination around coal-based thermal power plant in India. Environ Geol 51(3):409–420

Mandal A, Sengupta D (2003) Radioelemental study of Kolaghat thermal power plant, West Bengal, India—Possible environmental hazards. Environ Geol 44:180–187

Maraziotis EA (1985) Gamma activity of the fly ash from a Greek power plant and properties of fly ash in cement. Health Phys 49:302–307

Menon R, Raja P, Malpe D, Subramaniyam KSV, Balaram V (2011) Radioelemental characterization of fly ash from Chandrapur Super thermal power station, Maharashtra, India. Curr Sci 100:1880–1883

Mondal T, Sengupta D, Mandal A (2006) Natural Radioactivity of ash and coal in major thermal power plants of West Bengal, India. Curr Sci 91:1387–1393

Papastefanou C (2010) Escaping radioactivity from coal-fired power plants (CPPs) due to coal burning and the associated hazards: a review. J Environ Radioact

Parial K (2016) Integrated study for risk and vulnerability assessment using geochemical, Geophysical and geospatial analysis around Kolaghat thermal power plant, Eastern India, PhD. thesis, Indian Institute of Technology Kharagpur, India

Parial K, Guin R, Agrahari S, Sengupta D (2016) Monitoring of Radionuclide migration around Kolaghat thermal power plant, West Bengal, India. J Radioanal Nucl Chem 307(1):533–539

Parial K, Biswas A, Agrahari S, Sharma SP, Sengupta D (2015) Identification of contaminated zones using direct current (DC) resistivity surveys in and around ash ponds near Kolaghat thermal power plant, West Bengal, India. Int J Geol Earth Sci 1(2)

Seredin VV, Dai S (2012) Coal deposits as potential alternative sources for lantanidesnd yttrium 94:67–93

Singh S, Ram LC, Masto RE, Verma SK (2011) A comparative evaluation of minerals and trace elements in the ashes from lignite, coal refuse, and biomass fired power plants. Int J Coal Geol 87:112–120

Singh AK, Jojo PJ, Khan AJ, Prasad R, Ramchandran TV (1997) Calibration of track detectors and measurement of radon exhalation rate from solid samples. Radiat Prot Environ 3:129–133

Siotis I, Wrixon AB (1984) Radiological consequences of the use of fly ash in building materials in Greece. Radiat Prot Dosim 7:101–105

Stranden E (1983) Assessment of radiological impact of using fly ash in cement. Health Phys 44:145–151

Telford WM, Geldart LP, Sheriff RE (1990) Applied geophysics, 2nd edn. Cambridge University Press

Tufail M, Mirza SM, Chughati MK, Ahmad N, Khan HA (1991) Preliminary measurements of exhalation rates and diffusion coefficients for radon in cements. Nucl Tracks Radiat Meas 19:427–428

Ulbak K, Jonnassen J, Backmark K (1984) Radon exhalation from samples of concrete with different porosities and fly ash additives. Radiat. Prot. Dosim. 7:45–51

UNSCEAR (1993) Sources and effects of ionizing radiation. United Nations scientific committee on the effect of atomic radiation, United Nations, New York

UNSCEAR (2000) Sources and effects of ionizing radiation. United Nations scientific committee on the effect of atomic radiation, United Nations, New York

Vijayan V, Behera SN (1999) Studies on natural radioactivity in coal ash. In: Mishra PC, Naik A (eds) Environmental management in coal mining and thermal power plants. Technoscience, Jaipur, pp 453–456
Zielinski RA, Budahn JR (1986) Radionuclides in coal and coal combustion waste products: characterization of coal and coal combustion products from a coal-fired power plant. USGS Open File Report, pp 98–342

Forecasting Solid Waste Generation Rates

Sudha Goel, Ved Prakash Ranjan, Biswadwip Bardhan and Tumpa Hazra

Abstract Design of efficient solid waste management (SWM) systems requires accurate estimates of waste generation rates. Waste generation rates can be measured by direct sampling and several methods like database mining and sample surveys have been used in different locations. Another approach is to use different types of models to predict or forecast waste generation rates based on knowledge of impacting factors. These models include statistical models like multiple linear regression, econometric models, system dynamics methods, time-series analysis, factor analysis and Geographical Information Systems (GIS)-based methods. More non-conventional methods like artificial neural networks (ANN), fuzzy logic (FL) and support vector machine (SVM) methods are becoming popular. A brief literature review of these methods used by researchers for forecasting solid waste generation rates is provided in this chapter.

Keywords Statistical · Econometric · Regression · Sampling · Soft computing · Solid waste generation models · Economic solid waste management

Introduction

The objective of Solid Waste Management (SWM) is to control the functions of generation, collection, separation, transfer and transport, treatment and disposal of SW so that there are no adverse effects on either public health or the environment. Thus, efficient management of municipal solid waste (MSW) has become a major concern for urban local bodies (ULBs) throughout the world, especially the developing world. Several factors need to be accounted for in MSW management such as economic, technical, environmental, legislative, and political. Successful

B. Bardhan · T. Hazra
Civil Engineering Department, Jadavpur University, Kolkata, India

S. Goel (✉) · V.P. Ranjan
Civil Engineering Department, IIT Kharagpur, Kharagpur, India
e-mail: sudhagoel@civil.iitkgp.ernet.in; sudhagiitkgp@gmail.com

D. Sengupta and S. Agrahari (eds.), *Modelling Trends in Solid and Hazardous Waste Management*, DOI 10.1007/978-981-10-2410-8_3

design of treatment facilities and implementation of SWM policies depends on reliable statistics regarding solid waste generation rates, composition, and characteristics. In general, data regarding MSW generation or collection are collected on a regular basis in middle- and high-income countries while data are extremely difficult to obtain in developing countries like India since routine monitoring and data collection are rarely done. In such cases, SW collection data are used to estimate generation rates. Since knowledge of SW generation rates is the basis of all SWM strategies, there has been growing interest and research activity in this area in the last few decades. A brief literature review about factors affecting solid waste (SW) generation rates, and models developed and used for forecasting SW generation rates is presented in this chapter.

Factors Affecting Solid Waste Generation Rates

SW generation rates have been increasing all over the world mainly due to increase in population accompanied by an increase in resource consumption, making accurate predictions for the future a real challenge. There are two major parameters that are used in the literature for SW generation rates: total SW generated in a given location (weight/unit time, i.e., tons/day or tons/year) and per capita SW generated in a given location (weight/person-unit time, i.e., kg/person-day). Total SW generated is directly proportionate to the population of an area (McBean and Fortin 1993; USEPA 1997) since total SW generation rate (kg/d) = total population (persons) * per capita SW generation rate (kg/person-d).

An example illustrating a simple method for forecasting solid waste generation rates for a country like India is shown in the following problem.

Problem According to the 1901 Census of India, total population of India in 1901 was $2.38 \text{ x}10^8$ persons, and the urban population was 2.57×10^7 persons. The latest Census (2011) data show a total population of 1.21×10^9 persons, and an urban population of 3.77×10^8 persons.[1] Assume exponential growth applies in all cases, i.e., $N = N_0 \exp(kt)$, where N = population at time t, persons, N_0 = population at time $t = 0$, persons; k = annual exponential growth rate (1/year), t = time, years.

a. Calculate growth rates for the total population and the urban population in India over this period.
b. Assuming the current growth rates will continue, predict the total and urban populations in India in 2030.
c. If per capita resource consumption rates in urban India are growing by 7 %, and assuming that per capita urban solid waste generation rates will

[1] http://www.censusindia.gov.in/2011census/population_enumeration.html

follow the same trend, what are the total urban solid waste generation rates (tons/year) in 2011, assuming the per capita generation rate is 0.5 kg/cap-d. Also predict urban SW generation rates in 2030.

Solution

a. Based on the equation for exponential growth,
   Annual exponential growth rate for total population = $k = \ln(N/N_0)/t$ = 0.0148 1/year or 1.48 %.
   Similarly, annual exponential growth rate for the urban population = k_u = $\ln(N/N_0)/t$ = 0.0244 1/year or 2.44 %.
b. Based on the growth rates calculated in part a,
   Total population in India in 2030 = 1.6×10^9
   Urban population in India in 2030 = 6.0×10^8
c. Total urban solid waste generated in 2011 = 0.5 kg/cap-d * $3.77 \times 10^8 = 6.88 \times 10^7$ tons/y
   For calculating total urban solid waste generated in 2030, we know that both terms in the previous equation are growing exponentially, i.e., per capita resource consumption rate and population.
   Therefore, total urban solid waste generation rate in 2030 = 0.5 kg/cap-d *exp(0.07 1/year*19 years) *3.77×10^8 *exp(0.0244*19)
   =1.89 * 6.0×10^8 kg/d
   =4.134×10^8 tons/y

Several social, economic, and demographic factors are known to influence SW generation rates and have been reviewed by many researchers (Beigl et al. 2008; Yousuf and Rahman 2007; Keser et al. 2012; Cherian and Jacob 2012; Mohee et al. 2015; Owusu-Sekyere et al. 2015; Hoornweg and Bhada-Tata 2012; Kansal 2002). Results of our literature survey are summarized here:

i. *Population:* Total SW generation rates are directly proportionate to the population of the area while per capita SW generation rates are affected by the size of the area and other factors described below (Hockett et al. 1995; Vesilind et al. 2002, Goel 2008).
ii. *Income:* Financial resources at the individual and higher scales, i.e., city, regional, or national scales were found to be the most important factors affecting per capita SW generation rates, and therefore total SW generation rates (IBRD-World Bank 1999; Viswanathan 2006; Matsunaga and Themelis 2002). Parameters like individual or household income, gross domestic product (GDP), or gross national product (GNP) were found to correlate directly with per capita SW generation rates (Khan and Burney 1989; Hockett et al. 1995; Wang and Nie 2001; Chung 2010; Buenrostro et al. 2001; Phuntso et al. 2009). Further, income and GDP also determine the

composition of SW. The most important factors in determining SW generation in California were employment and taxable transactions (USEPA 1997). Since income is directly related to the ability to spend and consume resources, factors affecting consumption patterns will also affect SW generation rates (Purcell and Magette 2009; Daskapoulos et al. 1998).

A major factor associated with GDP or GNP is the *degree of economic development* resulting in significant differences in total and per capita SW generation rates in low, middle, and higher income areas or countries. The percentage of commercial waste can be predicted based on employment by sector (Bach et al. 2004; Hockett et al. 1995; Gay et al. 1993), while building area, construction area, population, and costs associated with various construction activities are commonly used to predict generation of construction and demolition waste (Yost and Halstead 1996; Shi and Xu 2006; Bergsdal et al. 2007; Chung 2010)

iii. *Recycling rates, legislation and policy measures*: In a study in Seattle, USA during the period of 2000–2010, population and recycling rates were found to have a significant impact on total SW generation rates. The commercial sector generated the greatest amount of SW (48 %) followed by single-family homes (30 %), self-hauling (12 %) and multiple family homes (10 %).[2] Policy and economic incentives were found to be driving waste reduction in villages in Prespa Park, a transboundary region of the Balkans with territory in Albania, Greece, and Macedonia (Grazhdani 2016). Recycling rates in this study increased where people were charged in proportion to the waste generated, where curbside recycling services were provided and where the number of drop-off recycling centers/1000 people was greater. Campaigns to sensitize people about the importance of waste reduction were found to be effective in improving recycling rates.

iv. *Education:* People with higher education levels generated less waste (Grazhdani 2016).

v. *Household or family size:* In a survey of 336 households in Kathmandu, Nepal, during 2007, total SW generation rate increased with increase in family size while per capita SW generation rates decreased with increase in family size (Dangi et al. 2011). Similar results were found in a sample survey of 1036 households in Dublin, Ireland in 1992 (Dennison et al. 1996). In a survey of 55 households in Chittagong City, Bangladesh, total MSW generation was directly proportionate to family size with an r value of 0.8483 (Salam et al. 2012).

vi. *Seasonal, geographic, or climatic factors:* Data from 20 urban areas of Iran showed clear changes in SW generation rates with season and geography (altitude of city) (Azadi and Karimi-Jashni 2016). Mean per capita SW generation rates were calculated and found to be correlated to population to

[2]http://www.seattle.gov/util/cs/groups/public/@spu/@garbage/documents/webcontent/02_015204.pdf

the greatest extent (correlation coefficient of 0.61) followed by maximum temperature, and therefore seasons (correlation coefficient of 0.48). Altitude and collection frequency did not have a major impact on MSW generation in this study. Similar seasonal changes in SW generation rates are mainly due to increase in tourist populations during the period from April to October were reported for the Island of Crete (Gidarakos et al. 2006). Seasonal effects were noted in a household survey in Chihuahua, Mexico, where total SW generation rates were highest in April, followed by August and January (Gómez et al. 2009). The increase in waste generation was attributed to greater consumption of packaged food and beverages, greater wastage of unconsumed food, and greater use/waste of vegetable fibers used in air conditioners (water coolers). A review of MSW generation in Dhaka city reports that there was a significant difference in the amount of waste arriving at the disposal site in the dry season (970 tons/d) versus the wet season (1420 tons/d) (Yousuf and Rahman 2007). Seasonal effects were clear for the Helsinki region, Finland in the database analyzed by Korhonen and Kaila (2015). Data analysis of MSW collections in the city of Sapporo, Hokkaido, Japan also show clear seasonal trends in the amounts of SW generated, and therefore collected (Matsuto and Tanaka 1993).

vii. *Collection frequency:* Collection frequency has a direct positive (R = 0.42) impact on the total amount of SW collected and no impact on per capita SW generation rates (Azadi and Karimi-Jashni 2016).

viii. *Others:* The amount of concrete debris generated was estimated based on cement production and building area (Shi and Xu 2006).

Methods for estimating MSW generation rates are generally based on the amount of waste collected by the municipal or administrative authorities and are based on either load-count analysis or weighing balance measurements of waste collected (King and Murphy 1996; Gidarakos et al. 2006). There are some reports in the literature which include generation data from individual sources, but these are short-term surveys and are reviewed in the section on sample surveys.

Bibliometric Analysis of Models Used for Forecasting Solid Waste Generation Rates

A bibliometric survey using SCOPUS was done to understand the level of interest and research activity in the area of solid waste management and models used for predicting solid waste generation rates. Using the keywords "solid waste management model" and restricting the search to English language publications, 4712 publications were found. Within these publications, only articles and reviews accounted for 3714 publications. In a recent review by Pires et al. (2011) about SWM systems in European countries, it was found that Life Cycle Assessment or Inventory was the most popular topic with 72 out of 218 articles published in the

Table 1 Top 10 countries and institutions based on number of publications

Rank	Country	Number of publications	Institution	Number of publications
1	United States	267	Danmarks Tekniske Universitet	18
2	China	121	University of Regina	17
3	Canada	74	North Carolina State University	14
4	Italy	58	Universiti Kebangsaan Malaysia	12
5	United Kingdom	58	Imperial College London	12
6	India	56	University of Tehran	12
7	Japan	49	Tsinghua University	12
8	Spain	45	Tongji University	12
9	Brazil	35	National Cheng Kung University	11
10	Malaysia	31	Chinese Academy of Sciences	11

Keywords Solid waste generation models

European Union (EU), followed by sustainability assessment (31) and management information systems (26) amongst the system assessment tools. Of the systems engineering models, there was far less interest in EU (only 38 out of 218 articles) of which cost–benefit analysis accounted for 12, followed by forecasting models (the topic of this review) which had only 10 publications.

Another search was done with the keywords "solid waste generation model" in SCOPUS. Restricting the search to English language publications resulted in a list of 1111 publications which include articles, reviews, conference papers, reports, books, etc.[3] Further analysis of the data from SCOPUS and information regarding the top 10 in the categories of country of publication, institution, subject, and source title are summarized in Tables 1 and 2. Publication activity in this area is shown in Fig. 1 for the period 1972–2016. In similar analyses for shorter time periods: 1991–2010 (Ma et al. 2011) and 1997–2011 (Yang et al. 2013), exponential growth in research publications in this subject area was noted in both studies.

Models/Methods Used for Forecasting Solid Waste Generation Rates

Various models have been developed for forecasting SW generation rates precisely. Salhofer (2001) has classified generation forecasting models in two major groups (Beigl et al. 2008):

- *Factor models* that use factors like consumption or utilization to describe the waste generation process.

[3]Note: Not all publications listed in SCOPUS are pertinent to the topic. These numbers are indicative but not accurate.

Table 2 Top 10 subject areas and sources based on number of publications

Rank	Subject	Number of publications	Source or Journals	Number of publications
1	Environmental Science	629	Waste management	103
2	Engineering	285	Waste Management And Research	63
3	Energy	209	Resources Conservation And Recycling	34
4	Earth and Planetary Sciences	124	Environmental Science And Technology	20
5	Chemical Engineering	99	Water Science And Technology	18
6	Medicine	79	Journal Of Environmental Management	15
7	Chemistry	65	Advanced Materials Research	14
8	Agricultural and Biological Sciences	43	Applied Energy	12
9	Materials Science	42	Journal Of Solid Waste Technology And Management	12
10	Social Sciences	42	Journal Of Environmental Engineering	11

Keywords Solid waste generation models

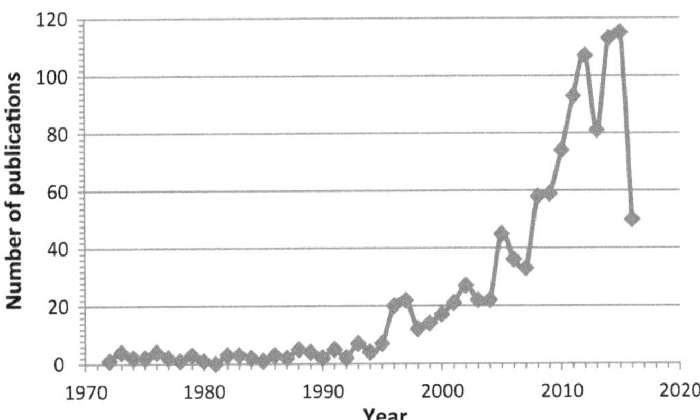

Fig. 1 Number of publications pertaining to "solid waste generation model". *Source of data* SCOPUS, downloaded on 10 Aug 2016

- *Input–output models* based on an accounting of the flow of materials into or out of a waste generation system.

Most of the models are based on deterministic methods or trend analysis without considering the dynamic properties of municipal SW generation. Factors may be

Table 3 Methods and models used for forecasting SW generation rates

Model/method used	Number of publications
Conventional methods/models	
Database mining/collection	7
Sample survey method	12
Linear regression	13
Econometric models	4
System dynamics	15
Time series analysis	5
Factor analysis	9
Geographic Information Systems (GIS)	5
Non-conventional methods/models	
Artificial neural networks (ANN)	16
"Fuzzy" methods	8
Adaptive Network-based Fuzzy Inference System (ANFIS)	4
"Gray" methods	5
Support Vector Machine (SVM)	3
Total publications	106

constant or fixed over time, i.e., they may be static models or they may be dynamic models where the factors change with time. Forecasting with traditional statistical forecasting methods requires collection of socio-economic, demographic, and environmental data (Dyson and Chang 2005). In many cases, municipalities do not have sufficient finances and human resources to create a complete database of SW quantity and composition on a long-term basis. Short-term forecasting of future SW generation rates is necessary for better design of storage, collection, and transfer systems (Matsuto and Tanaka 1993) while long-term forecasting is needed for selecting appropriate waste treatment technologies or selecting landfill sites or understanding the impacts of new policies and initiatives. However, if attitudes of waste generators change or new/revised regulations regarding waste management are implemented during the forecast period, long-term forecasting can go awry.

Forecasting methods have been categorized as conventional or non-conventional methods/models for this review. A list of the methods and models reviewed in this chapter and the number of publications accessed is shown in Table 3.

Conventional Methods/Models Used for Solid Waste Generation Forecasting

Several conventional methods have been used for forecasting SW generation rates and were listed in Table 3 along with the number of publications that were available for each of the methods. Reviews of these methods are also available and brief descriptions are included in this section (Beigl et al. 2008; Cherian and Jacob 2012; Chung 2010; Pires et al. 2011).

Data Collection and Database Mining

In many high- and middle-income countries, monitoring and collection of solid waste generation data is done on a regular basis and is termed 'data collection' here. If this practice has been followed in several locations and over a long period of time, it has resulted in large amounts of data and in some cases, these databases are available to researchers. Many researchers have been able to access these databases and use pertinent data for forecasting future trends and determining factors that affect SW generation rates. Database mining is the development of algorithms for extracting pertinent data from large databases for further analysis. Two studies noted in Table 4 are based on database mining (Korhonen and Kaila 2015; Lebersorgher and Beigl 2011). The remaining studies in this Table are based on data collection from the literature or administrative database/documents (secondary sources).

Sample Survey Method (SSM)

This is the simplest and most common method used for estimating SW generation rates. SW generation studies based on sample surveys are summarized in Table 5. The first step in this method is to determine a representative sample size and the number of strata to be sampled (King and Murphy 1996). Then primary data are collected by personal interviews or surveys or secondary data like Census or municipal data are used directly. In general, sample-based surveys are stratified based on criteria like income, season, sources of waste, and several other factors. A summary of sample-based studies accessed for this review is provided in Table 5.

Most of the studies (9 out of 12) reviewed here were based on house-to-house sampling for a short period of time ranging from 1 day to 2 weeks. Interview or questionnaire-based surveys for collecting socioeconomic and demographic data were accompanied by waste collection surveys to determine waste generation rates. In general, one bag for total waste collected in one day or two bags for kitchen or wet waste and dry waste; toilet waste and other waste were provided to each household identified for the survey. In some studies, one bag was provided per household and collected after 24 h (Phuntsho et al. 2009; Salam et al. 2012). The waste contents were then sorted to determine individual waste components. In other studies, sampling was carried out for a period of 8 days (Gómez et al. 2009), 10 days (Grazhdani 2016) to 2 weeks (Thanh et al. 2011; Dangi et al. 2011), and the waste bags collected every 24 h for the duration of the survey. Grazhdani (2016) and her survey team collected 20 samples/village for 10 days/season for 4 seasons. Daily production of waste and recyclables was monitored and 1 m^3 of waste in every village was collected, segregated, analyzed, and individual components weighed.

Dennison et al. (1996) conducted a household survey in Dublin, Ireland, and found that more affluent households generated more waste and total waste produced was directly proportionate to the number of people in the household. These findings

Table 4 Publications based on data collection or data mining for forecasting SW generation rates

	References	Area	Period of study	Number of data points or sets	Parameter(s) measured
1	Korhonen and Kaila (2015)	Helsinki region, Finland	2013	27,865	Household waste in 6 m^3 container
2	Bridgwater (1986)	UK	1930–1982	53	Annual waste generation and composition data
3	Purcell and Magette (2009)	Dublin, Ireland	2006	322 electoral districts, 2261 commercial establishments	Census, electoral and other administrative databases
4	Lebersorgher and Beigl (2011)	Styria, Austria	1991–2008	116 socioeconomic and demographic parameters for 42 municipalities	
5	Khan and Burney (1989)	Global data	–	28 cities	Wages, temperature, population and population density, waste composition, and household size
6	Niessen and Alsobrook (1972)	New Jersey and New York states	2000	41 datasets	Waste composition
7	Hibiki and Shimane (2006)	Kanto, Japan	1995–2002	441 municipalities	Population, area, income, number of establishments, weight of refuse and recyclables, collection frequency, price of trash bag

have been confirmed in many other studies. Phuntsho et al. (2009) used the sample survey method to calculate SW generation rates for Bhutan. In the largest survey conducted in Bhutan for 10 cities, the average SW generation rate was found to be 0.53 kg/cap-d with residential sources contributing the maximum amount of waste (47 %). No correlation between household income and per capita SW generation rate was found. A household survey was conducted in the dry (100 households) and rainy seasons (60 households) in the city of Can Tho, Vietnam, and its surrounding region to quantify and characterize residential solid waste (Thanh et al. 2010). No significant differences in waste generation were found for the total waste generated/household in the wet and dry seasons. The authors attributed this to the lack of difference in meteorological characteristics for the city. Waste composition was affected by income while kitchen waste (a subcategory of total waste) was the only category which showed no impact of income. Generation of plastic, food, and total SW was well correlated to household size while paper and textile waste generation were correlated well with levels of urbanization. Buenrostro et al.

Table 5 Summary of Sample survey studies for estimating or predicting SW generation rates

	References	Area	Period of study	Classification or strata	Criteria for classification	Total number of samples	Parameter(s) measured
1	Dennison et al. (1996)	Dublin, Ireland	1992	Not classified	Not classified	1036 households	Socioeconomic, demographic and waste composition
2	King and Murphy (1996)	Broward Country, Florida, USA	1990–1991	64	Districts (8), collection frequency (2 times/week), data collection frequency (1 time/quarter)	64; one truckload/strata	Number of truckloads, weight per truckload, number of households/route of one truck; only residential waste
3	Phuntso et al. (2009)	Bhutan	2007–08	10 towns	Sources of waste: residential, commercial, markets, offices, schools and institutions	11,068 households	Waste generated/household, number of persons/household or institution, household income
4	Benítez et al. (2008)	Mexicali, Mexico	2005 (?)	3	Three stages (or phases) in three years	181 households	Weight, composition, consumption habits, number of children and person/household, education, income
5	Thanh et al. (2010)	Can Tho city, Vietnam	2009	2	Season	100 households	Household solid weight, composition, population density, number of person/household, education, income, level of urbanization,
6	Buenrostro et al. (2001)	Morelia, Mexico	1998	3	Socioeconomic level	243 households	Weight, number of person/household or institution, education, income
7	Salam et al. (2012)	Chittagong City, Bangladesh	2009	5	Socioeconomic level	55 households	Weight, composition, income
8	Ojeda-Benıtez et al. (2008)	Colonias, Mexicali, Mexico	1999–2000	Not classified	Not classified	160 households	Weight, composition, number of children and person/household

(continued)

Table 5 (continued)

	References	Area	Period of study	Classification or strata	Criteria for classification	Total number of samples	Parameter(s) measured
9	Grazhdani (2016)	Prespa Park	2014	Not classified	Not classified	410 households	Weight, composition, number of person/household, income, education, ages of residents and building, travel time to work, access to curbside recycling and drop-off centers
10	Gómez et al. (2009)	Chihuahua, Mexico	2006–07	3 income levels and 3 seasons	Season, socioeconomic level	1687 samples	Weight, composition, consumption habits, number of children and person/household, education, income
11	Dangi et al. (2011)	Kathmandu, Nepal	2007	4	Income	336 households	Weight, composition, income, household size
12	Gidarakos et al. (2006)	Island of Crete	2003–04	4 phases, 7 weeks each	Population: permanent and tourist, season	Vehicle load counts	Weight, composition

(2001), Benítez et al. (2008) conducted random household sampling surveys stratified on the basis of household income in Morelia, Mexico.

King and Murphy (1996) designed a sample survey to estimate the amount of solid waste to be collected on an annual basis using curbside collection in the unincorporated residential part of Broward County, Florida, USA. Their estimates of waste generation were based on number of truckloads picked by each truck on one route. They also included effects of collection day and frequency (2 times/week) and data were collected and analyzed on a quarterly basis. An interesting effect noted was that waste generated after weekends was greater than in the latter part of the week. In another study for estimating SW generation rates in Crete, Gidarakos et al. (2006) made several assumptions: for populations < 10,000, per capita SW generation rate of 0.8 kg/cap-d, for populations > 10,000, per capita SW generation rate of 1.0 kg/cap-d and for large municipalities, per capita SW generation rate of 1.2 kg/cap-d. For estimating waste generation by tourists, they determined total number of tourist beds, beds in use and a per capita SW generation rate of 1.2 kg/cap-d. As part of the sampling survey which was conducted for 28 weeks (5 days/week), vehicle loads were counted and the total SW collected was estimated.

Multiple Linear Regression (MLR) Method

Linear regression analysis of data has been used for more than 200 years old with the earliest publication of the method in 1805 by LeGendre (Abdi 2007). SW generation rates can be forecast based on either one variable (single regression) or several variables (multiple linear regression). Since SW generation forecasting is dependent on several factors, multiple linear regression method is commonly used. A list of publications accessed for this review is provided in Table 6.

Table 6 Multiple linear regression models used for predicting SW generation rates in different locations

	Location of application	References
1	Fars, Iran	Azadi and Karimi-Jashni (2016)
2	New Delhi, India	Vivekananda and Nema (2014)
3	Beijing, China	Wei et al. (2013)
4	Cities - global data	Khan and Burney (1984)
5	Mexicali, Baja California, Mexico	Benítez et al. (2008)
6	Can Tho city, Mekong Delta, Vietnam	Thanh et al. (2010)
7	Morelia, Mexico	Buenrostro et al. (2001)
8	Beijing Satellite Towns, China	Ying et al. (2011)
9	Beijing, China	Wang and Nie (2001)
10	Waterloo, Ontario, Canada	McBean and Fortin (1993)
11	Ankara, Turkey	Keser et al. (2012)
12	Washington DC, USA	Joutz (1996)
13	Vienna, Austria	Bach et al. (2004)

Khan and Burney (1989) used multiple linear regressions to quantify the impact of an individual's income, population, population density, number of residents per house, and GDP on the composition of the solid waste. Grossman et al. (1974) used population, income level, and the dwelling unit size for predicting SW generation by linear regression analysis. Hockett et al. (1995) developed a linear regression model to determine the effects of demographic, economic, and structural factors on per capita SW generation rates. They also evaluated the specific contribution of retail sales to the generation of waste in the southeastern region of the U.S.A. They concluded that retail sales and waste disposal fees are important factors that determine waste generation rates while manufacturing, construction, personal income, and degree of urbanization were not important. Benítez et al. (2008) in their linear regression model used composition, consumption, education, household income, and number of residents per house to determine SW generation rates. Data in this study was collected through household sampling surveys. They developed several multiple linear regression equations and found that the equation with 3 independent variables gave the best-fit with an R^2 of 0.51. Per capita SW generation was positively correlated to income and negatively correlated to education and number of persons in the household. Similarly, data generated by Buenrostro et al. (2001) in sample surveys was analyzed using multiple linear regression to predict future solid waste generation rates while accounting for various socioeconomic variables. For residential sources of waste, household income and size were significant factors while numbers of working hours were significant factors for non-residential sources of SW.

Thanh et al. (2010) developed a multiple linear regression model for Can Tho city, the capital city of the Mekong Delta region in southern Vietnam to evaluate the correlation between household size and household income with SW waste generation rates. In a survey of 542 municipalities, per capita SW generation rates in Styria, Austria were found to be negatively correlated to household size, illegal use of waste as fuel, and per capita municipal tax which is an indirect indicator of the per capita income of the municipality (Lebersorger et al. 2011). Multiple linear regression analysis of data for China showed that total MSW generation rates were directly proportionate to the urban population and GDP with an r^2 value of 0.9858 (Wang and Nie 2001).

Econometric Forecasting Methods (EFM)

"An econometric model is one of the tools economists use to forecast future developments in the economy. In the simplest terms, econometricians measure past relationships among such variables as consumer spending, household income, tax and interest rates, employment, and the like, and then try to forecast how changes in some variables will affect the future course of others" Hymans (2008). In SWM, economic factors are of major significance and forecasting generation rates must account for these factors and uncertainties associated with them. When the economic impact of recycling is phenomenal as is the case in developing countries like

Table 7 Summary of econometric models used for predicting SW generation rates

	References	Location of application	Parameter used for prediction of SW generation rates
1	Gay et al. (1993)	King and adjacent counties, Washington State, USA	Economic sales data
2	Daskalopoulos et al. (1998)	EU (1980–93), USA (1960–1993)	Total consumer expenditure and GDP
3	Chunsheng (2009)	Jinan, China, 1999–2008	Household consumer expenditure
4	Bruvoll and Ibenholt (1997)	Norway, industrial waste	Input of raw materials, production of goods and other important economic variables

India, socioeconomic and demographic factors acquire even greater importance and must be included in SW generation rate forecasts. Applications of econometric models to generate disaggregated predictions of residential solid waste loads were presented by Grossman et al. (1974). Other methods of forecasting SW generation rates based on economic indicators have been presented or reviewed in IBRD-World Bank (1999), Daskalopoulos et al. (1998a), Matsunaga and Themelis (2002), Chung and Poon (1998), Jena and Goel (2014), Goel (2008), Cherian and Jacob (2012), and Viswanathan (2006). Some publications reviewed here are listed in Table 7.

Since direct measurements of SW generation rates are difficult in terms of financial and human resource inputs, several researchers have developed indirect methods of estimating SW generation rates and economic variables are considered one of the best ways of doing so. Gay et al. (1993) used economic sales data for King County, Washington state, USA to develop conversion factors for estimating SW generation rates. They validated their results using data from adjacent counties. Based on the concept that total MSW generation is correlated to GDP and population for a country, Daskalopoulos et al. (1998) developed a methodology/algorithm for estimating MSW generation for any country based on data for EU and USA. Inputs required were GDP, and population from which total consumer expenditure and related consumer expenditure were calculated. Based on previous data and these calculations, the total amount of MSW and MSW components can be predicted or estimated. Chunsheng (2009) conducted a similar study for Jinan in China. Bruvoll and Kaila (1997) developed an econometric model for Norway's industrial solid waste generation based on production rates or inputs of raw materials.

A bibliometric survey of the literature (English language only) on this topic showed 2958 results with the keywords: "economic solid waste management in SCOPUS" (as of 6 August 2016). Of these results, 1145 publications pertained to "models" and 92 publications were reviews. These preliminary bibliometric results are indicative of the relatively recent but growing interest in this topic as shown in

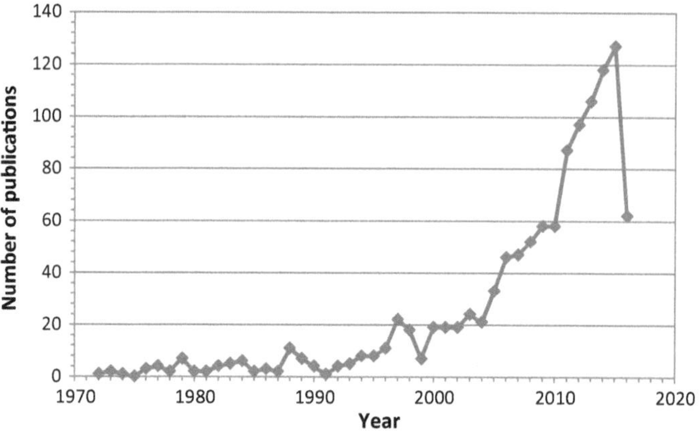

Fig. 2 Publications in the area of econometric models applied to solid waste management. *Source of data* SCOPUS

Fig. 2 by the number of publications listed in SCOPUS from 1972 to 2016. Top 10 results (number of publications) associated with other bibliometric parameters such as country of publication, institution, subject and source are summarized in Tables 8 and 9.

System Dynamics Method (SDM)

System dynamics method enables researchers to examine interactions among demographic, social, economic, environmental, and decision-making factors. "System dynamics is a well-established methodology for studying and managing complex feedback systems" (Dyson and Chang 2005). It requires constructing unique "causal loop diagrams" or "stock and flow diagrams" like those in the software STELLA. The advantage of system dynamic models is that they can use past data to simulate results for the future, if consistent data are available. Several applications of SDM have been published and some are summarized in Table 10.

Chen et al. (2012) used System Dynamics (SD) model for MSW management to explore whether the present waste disposal capacity can meet the increasing amount of waste generation for Singapore. The parameters which they considered were the initial MSW generation, disposal and composition of the MSW, population, household size, GDP, socioeconomic factors and control target (SGP 2012), landfill capacity, incineration capacity, recycle composition and collection rate from National Environmental Agency (NEA). Pai et al. (2014) used System Dynamics methodology to study the impact of total population, population growth and decline, amount of domestic waste generated and collected and collection volume of hospital waste on the amount of solid waste generation. However, the results of this model were not validated with real data.

Table 8 Top 10 countries and institutions based on number of publications

Rank	Country	Number of publications	Institution	Number of publications
1	United States	228	University of Regina	49
2	China	128	North China Electric Power University	42
3	Canada	98	National Cheng Kung University	22
4	Italy	76	Danmarks Tekniske Universitet	14
5	United Kingdom	54	Universidade Federal do Rio de Janeiro	12
6	India	50	Universiti Kebangsaan Malaysia	12
7	Brazil	46	The Royal Institute of Technology KTH	12
8	Sweden	46	Imperial College London	12
9	Spain	45	Universiti Teknologi Malaysia	11
10	Malaysia	38	Peking University	9

Keywords Economic solid waste management

Time Series Analysis (TSA)

Time series forecasting (TSA) is considered to be the most accurate method for predicting short-term variations like assessment of seasonal effects. "There are two main goals of time series analysis: (a) identifying the nature of the phenomenon represented by the sequence of observations, and (b) forecasting, i.e., predicting future values of the time series variable".[4] A list of publications pertaining to time series analysis for predicting SW generation rates is provided in Table 11.

Estimation techniques like S-, cubic, quadratic, linear curve, etc., exponential smoothing, Box and Jenkins method, and AutoRegressive Integrated Moving Average (ARIMA) models have been used for forecasting SWM generation rates (Chung 2010). Since predictions or forecasts are to be made on the basis of "time", the models can be run for annual, monthly or even daily intervals. According to Chung (2010), seasonal ARIMA and nonlinear techniques produced good results in terms of forecasting accuracy and mean relative error when daily and monthly SWM data were collected and analyzed for seasonal and other impacts (Navarro-Esbrí et al. 2002; Matsuto and Tanaka 1993; Katsamaki et al. 1998). Bridgwater (1986) used the S-curve technique for predicting waste generation rates and composition for up to 50 years.

[4]http://documents.software.dell.com/Statistics/Textbook/Time-Series-Analysis

Table 9 Top 10 subjects and sources based on number of publications

Rank	Subject	Number of publications	Source	Number of publications
1	Environmental Science	811	Waste Management	145
2	Engineering	185	Resources Conservation And Recycling	94
3	Energy	159	Waste Management And Research	75
4	Earth and Planetary Sciences	122	Journal Of Environmental Management	32
5	Medicine	113	Journal Of Cleaner Production	20
6	Economics, Econometrics and Finance	97	Water Science And Technology	19
7	Social Sciences	85	Journal Of The Air And Waste Management Association	16
8	Agricultural and Biological Sciences	67	Bioresource Technology	15
9	Chemical Engineering	60	Wit Transactions On Ecology And The Environment	15
10	Computer Science	43	Environmental Monitoring And Assessment	11

Keywords Economic solid waste management

Table 10 System dynamics models used for predicting SW generation rates in different locations

	Location of application	References
1	New York State	Mashayekhi (1993)
2	Chennai	Sudhir et al. (1997)
3	Berlin	Karavezyris et al. (2002)
4	San Antonio, Texas, USA	Dyson and Chang (2005)
5	Dhaka	Sufian and Bala (2007)
6	Newark, USA	Kollikkathara et al. (2010)
7	Singapore	Chen et al. (2012)
8	UK	Al-Khatib et al. (2015)
9	New Delhi	Vivekananda and Nema (2014)
10	Palestine (Israel)	Eleyan et al. (2013)
11	Hong Kong (China)	Wang et al. (2013)
12	Karnatka, India	Pai et al. (2014)
13	Kolkata, India	Dasgupta et al. (2016)
14	Jakarta, Indonesia	Chaerul et al. (2008)
15	Delhi, India	Ahmad (2012)

Table 11 List of locations and references where TSA was applied

	Location of application	References
1	Caruaru, Brazil	Neto et al. (2016)
2	Castellón, Spain	Navarro-Esbrí et al. (2002)
3	Chania, Greece	Katsamaki et al. (1998)
4	Tainan, Taiwan	Chang and Lin (1997)
5	Arusha City, Tanzania	Mwenda et al. (2014)

Time series analysis has several advantages (Chung 2010):

i. It is flexible, i.e., waste data measured at any meaningful interval can be used;
ii. Very few parameters are needed. In many cases, data for two variables only are needed: time and the key variable to be predicted.

However, TSA has some serious limitations as well since it does not include other important variables and their interactions. Predictions made with time series analysis may also lack general application and theoretical value, i.e., they do not always contribute to a fundamental understanding of the phenomenon.

"With time series, we can allow the dependent variable to be influenced by not only the current values of the independent variables, but past values as well. We can even include past values of the dependent variable. If the present can be modeled using only the past values of the independent variables, we can start to make very appropriate forecasting predictions."[5] Further, time series models require large amounts of data. For example, 50–100 equally spaced observations are required for ARIMA (Granger 1989). An ARIMA model is a combination of Autoregressive (AR) model which shows relationships between past and present values, a random value and a Moving Average (MA) model which shows that present values are related to past residuals. Accuracy of ARIMA methods depends on the amount of data available for formulating the waste trend and for validating a predicted trend. Thus, if only annual waste data are available, then insufficient data can limit application of TSA.

Factor Models (FM)

Factor models are based on identification of causal factors, their inter-relationships and describing these relationships in mathematical terms so as to predict future outcomes. For forecasting SW generation rates, this means identification of various social, economic, and demographic factors and other explanatory variables. Thus, factor models are more theoretically sound compared to the time series approach (Chung 2010). The stocks and flows model used by Bergsdal et al. (2007) are a specific form of factor models and have been used for forecasting construction waste generation rates. With suitable modifications, it can be used to forecast

[5]https://rpubs.com/ryankelly/tsa3

Table 12 List of locations and references where Factor models were applied

	Location of application	References
1	Xiamen Island, China	Xu et al. (2016)
2	Harbin, China	Chu et al. (2016)
3	Xiamen, China	Zhang et al. (2015)
4	Ankara, Turkey	Keser et al. (2012)
5	Tehran, Iran	Abdoli et al. (2011)
6	Nablus, Palestine	Arafat and Arafat (2011)
7	Dublin, Ireland	Purcell and Magette (2009)
8	Wisconsin, United States	Rhyner and Green (1988)
9	Jeddah, Saudi Arabia	Khan and Burney (1984)

commercial and industrial waste generation rates. Identification of valid explanatory variables is the most challenging work for Factor models. "This is often due to difficulties in including theoretically valid but hard-to-measure variables, such as the level of environmental awareness of the population, waste management literacy of the population and forms of waste policies deployed" (Chung 2010). A list of published studies based on Factor models is provided in Table 12.

Geographical Information System (GIS) Based Method

Total SW generation rate is not sufficient information for preparing a SWM plan. Along with the quantity of waste generated, its spatial distribution must also be known. With this information, it is possible to develop waste management strategies for locations where collection is most needed. Some publications based on GIS applications for estimating solid waste generation rates are listed in Table 13. Karadimas and Loumos (2008) used a geospatial database regarding population densities, commercial activities, road characteristics, etc., integrated in GIS environment to forecast MSW generation for the Municipality of Athens. Alsamawi et al. (2009) used GIS to forecast MSW generation rate for Baghdad considering population density and GDP. Purcell and Magette (2009) used GIS method to forecast household and commercial biodegradable waste generation considering socioeconomic conditions, household size, and existing waste generation rate for the Dublin region.

Non-conventional Models Used for Generation Forecasting

Data-driven models rely on input–output data without the need for complete perception of the complex processes that they are modeling. Three data-driven models are Artificial Neural Network (ANN), Fuzzy Logic-based model and Adaptive Neuro-Fuzzy Inference System (ANFIS) models. Soft computing techniques that have been successfully employed to solve the problems related to SW generation predictions are:

Table 13 List of locations and references where GIS was applied

	Location of application	References
1	Tennodai, Tsukuba-shi, Ibaraki, Japan	Akther et al. (2016)
2	Athens, Greece	Karadimas and Loumos (2008)
3	Nagpur, India	Gautam and Kumar (2005)
4	Baghdad, Iraq	Alsamawi et al. (2009)
5	Dublin, Ireland	Purcell and Magette (2009)

- Artificial Neural Network (ANN) method
- Fuzzy Logic (FL) Method
  - Gray Dynamic Modeling (GDM) approach
  - Adaptive Neuro-Fuzzy Inference Systems (ANFIS)
- Support Vector Machine (SVM) method

Artificial Neural Networks (ANN)

Complex, nonlinear events can be modeled with ANN which can make it useful for modeling Municipal Solid Waste Management (MSWM) Systems (Liu et al. 2002; Chi et al. 2005; Shu et al. 2006). References to some of these studies are provided in Table 14. Jalili and Noori (2008) used feedforward neural network to predict the weekly SW generation in Mashhad, Iran. Sudhir Kumar et al. (2011) used ANN

Table 14 List of locations and references where ANN models were applied

	Location of application	References
1	Fars, Iran	Azadi and Karimi-Jashni (2016)
2	Langkawi Island, Malaysia	Shamshiry et al. (2014)
3	Belgrade, Serbia	Antanasijević et al. (2013)
4	Tehran, Iran	Ali Abdoli et al. (2012)
5	Saqqez city, Kurdistan Province	Shahabi et al. (2012)
6	Malaysia	Shamshiry et al. (2011)
7	Mashhad, Iran	Jalili and Noori (2008)
8	Pune, India	Srivastava and Nema (2005)
9	Vijayawada, India	Kumar et al. (2005)
10	Wien, Austria	Wieland et al. (2002)
11	Eluru, A.P, India	Sudhir et al. (2011)
12	Taiwan, China	Shu et al. (2006)
13	Gujarat, India	Patel and Meka (2013)
14	Tehran, Iran	Noori et al. (2010)
15	Anhui, China	Liu et al. (2002)
16	Tokyo, Japan	Antanasijevic et al. (2013)

model to predict SW generation in the Eluru city, Andhra Pradesh, India from 2010 to 2026. Tiwari et al. (2012) predicted industrial SW generation using ANN model and then compared the value with results obtained from ANFIS model. Azadi and Karimi-Jashni (2016) used ANN and MLR to predict SW generation rates in Fars region of Iran.

Though ANN-based models have been successfully used in SW prediction, Wieland et al. (2002) opined that a major shortcoming of ANN is that causal relationships between major system components cannot be determined, and therefore there is no improvement in the user's understanding of the process. Another problem is that outputs in ANN-based models are based on the inputs. In cases where the opposite is needed, i.e., deriving inputs leading to a given output, neural networks cannot be used (Tiwari et al. (2012). However, algorithms for deriving qualitative rules from these models can be developed (Wieland et al. 2002).

Fuzzy Logic (FL)

Methods like ANN using input–output data require a complete database. In many cases, administrative authorities do not have a complete database of parameters affecting SW generation, SW quantity, and characteristics due to insufficient financial and human resources. Under these conditions, SW generation rate forecasting is dependent on subjective criteria and the weights of these factors are normally expressed in linguistic terms. Hence for such situations fuzzy logic is most suitable. Previous studies have indicated that the fuzzy set theory is more convenient and practical in accounting for uncertainties due to vagueness or fuzziness rather than randomness alone (Xi et al. 2008). Studies predicting solid waste generation rates based on fuzzy logic or fuzzy methods are listed in Table 15.

Adaptive Network-based Fuzzy Inference System (ANFIS)

Jang (1993) proposed a method called the Adaptive Network-based Fuzzy Inference System (ANFIS) by combining ANN and fuzzy systems. ANFIS models (Noori et al. 2009a) have become popular because:

Table 15 List of references and locations where Fuzzy Logic methods were applied

	Location of application	References
1	Athens, Greece	Karadimas and Orsoni (2006)
2	Tainan, Taiwan	Chen and Chang (2000)
3	Palermo, Italy	Raimondi et al. (1997)
4	Columbia, USA	Zeng and Trauth (2005)
5	Berlin, Germany	Karavezyris et al. (2002)
6	Berkeley, California	Jang (1993)
7	Baranagar, W. Bengal, India	Bardhan (2015)
8	Northern India	Khan and Farooqui (2012)

Table 16 List of references and locations where fuzzy-based methods like ANFIS were applied

	Location of application	References
1	Selangor, Malaysia	Younes et al. (2015)
2	Tamil Nadu, India	Arumugam et al. (2014)
3	Tehran, Iran	Noori et al. (2009a)
4	Durg-Bhilai Twin City, India	Tiwari et al. (2012)

i. they can be used for calibrating nonlinear relationships,
ii. they claim to be more efficient and accurate than conventional modeling techniques, and
iii. they can capture large amounts of nonlinear and noisy data, even when no physical or causal relationships are apparent (Beigl et al. 2008).
iv. Other associated benefits include improvement of model performance, and improvement in ability to predict outcomes through comprehensive bootstrapping operations (Openshaw and Openshaw 1997). Also, it can be used for short-, medium-, and long-term forecasting (Mordjaoui and Boudjema 2011).

For these reasons, ANFIS models have become popular for simulating environmental processes and several applications have been reported (Noori et al. 2009a; Tiwari et al. 2012) and are listed in Table 16.

Gray Dynamic Modeling (GDM)

Gray Dynamic Modeling is another method for forecasting issues in an uncertain environment. Applications of GDM to SW generation prediction are provided in Table 17. The Gray Fuzzy Integer Programming (GFIP) approach was developed by Huang et al. (2005) to deal with a solid waste management problem in Canada. Chen and Chang (2000) used gray fuzzy dynamic modeling for predicting SW generation rates in urban areas. Gray fuzzy dynamic modeling helps in minimizing differences between predicted values and observed values. Karadimas and Orsoni (2006) used fuzzy logic based modelin combination with GIS to forecast MSW generation rate in a small part of Athens. Oumarou et al. (2012) used fuzzy logic based modeling approach to predict MSW generation rates of Nigeria and recovery and recycling potential of SW.

Support Vector Machine (SVM) Method

"Models which have too many predictors and are of higher order polynomials begin to model the random noise in the data leading to a condition known as 'overfitting the model.'"[6] Overfitting during training, assignment of correct membership function, difficulty in determining network architecture, poor performance are some of the disadvantages of ANN, Fuzzy logic-based and ANFIS models. Support

---

[6]Wikipedia, 2016. Retrieved from https://en.wikipedia.org/wiki/Overfitting [9 August 2016].

Table 17 List of references and locations where gray dynamic models were applied

	Location of application	References
1	Thailand	Intharathirat et al. (2015)
2	Beijing, China	Ying et al. (2011)
3	Xiamen city, China	Xu et al. (2013)
4	Shangai city, China	Liu and Yu (2007)
5	Hypothetical scenario	Huang et al. (2005)

Table 18 List of references and locations where SVM methods were applied

	Location of application	References
1	Tehran, Iran	Abbasi et al. (2013)
2	Tehran, Iran	Abbasi et al. (2013)
3	Mashhad, Iran	Noori et al. (2009b)

Vector Machine (SVM) method is proposed for regression and classification purposes (Noori et al. 2009b). SVM is widely applicable and is less prone to overfitting unlike many other models. Further, SVM simultaneously minimizes error estimates and model dimensions unlike principal component analysis which only reduces the dimensionality of the model (Li et al. 2010). Therefore, several researchers now use SVM method to predict SW generation (Noori et al. 2009b; Abbasi et al. 2013). Table 18 summarizes some of the models that have been used so far for SW generation forecasting.

Evaluation of Model Performance

The performance of all models needs to be evaluated by comparing their outputs with real data. Many different parameters have been used to evaluate model performance. The most common parameters are the correlation coefficient (R), the coefficient of determination (R^2), root-mean-square error (RMSE), and mean absolute error (MAE). Definitions and equations for these are available in standard statistics textbooks.

Summary

Various factors affecting SW generation rates, models, and methods for predicting or forecasting SW generation rates have been reviewed in this chapter. Some of the models are static while others are dynamic. Some models are deterministic while others are probabilistic or stochastic. Models and methods have been categorized as conventional, i.e., those based on traditional methods and non-conventional for models based on soft computing methods and are described in this chapter.

References

Abbasi M, Abduli MA, Omidvar B, Baghvand A (2013) Forecasting municipal solid waste generation by hybrid support vector machine and partial least square model. Int J Environ Res 7(1):27–38

Abdi H (2007) The method of least squares. In: Salkind N (ed) Encyclopedia of measurement and statistics. Sage Publications, Thousand Oaks (CA)

Ahmad K (2012) A systems dynamics modeling of municipal solid waste management systems in Delhi. Int J Res Eng Technol 1(4):628–641

Akther A, Ahamed T, Takigawa T, Noguchi R (2016) GIS-based multi-criteria analysis for urban waste management. Nihon Enerugi Gakkaishi/J Japan Inst Energy 95(5):457–467

Ali Abdoli M, Falah Nezhad M, Salehi Sede R, Behboudian S (2012) Long-term forecasting of solid waste generation by the artificial neural networks. Environ Progress Sustain Energy 31:628–636

Al-Khatib IA, Eleyan D, Garfield J (2015) A system dynamics model to predict municipal waste generation and management costs in developing areas. J Solid Waste Technol Manage 41 (2):109–120

Alsamawi AA, Zboon ART, Alnakeeb A (2009) Estimation of Baghdad municipal solid waste generation rate. Eng Tech J 27:1

Antanasijevic D, Pocajt V, Popovic I, Redzic Nebojsa, Ristic M (2013) The forecasting of municipal waste generation using artificial neural networks and sustainability indicators. Sustain Sci 8(1):37–46

Arafat HA, Arafat AR (2011) Prediction of generation rate of municipal solid waste in Palestinian territories based on key factors modelling. Solid Waste Manage Environ Remed 425–440

Arumugam T, Parthiban L, Rangasamy P (2014) Two-phase anaerobic digestion model of a tannery solid waste: experimental investigation and modeling with ANFIS. Arabian J Sci Eng 40(2):279–288

Azadi Sama, Karimi-Jashni Ayoub (2016) Verifying the performance of artificial neural network and multiple linear regression in predicting the mean seasonal municipal solid waste generation rate: a case study of fars province, Iran. Waste Manage 48:14–23

Bach H, Mild A, Natter M, Weber A (2004) Combining socio-demographic and logistic factors to explain the generation and collection of waste paper. Resour Conserv Recy 41:65–73

Bardhan B (2015) Development of a fuzzy inference system based model to predict solid waste generation. M.Tech. Thesis, Civil Engineering, Jadavpur University

Beigl P, Lebersorger S, Salhofer S (2008) Modelling municipal solid waste generation: a review. Waste Manag 28(1):200–214

Benítez SO, Lozano-Olvera G, Morelos RA, Vega CA (2008) Mathematical modeling to predict residential solid waste generation. Waste Manag 28:S7–S13

Bergsdal H, Bohne RA, Brattebø H (2007) Projection of construction and demolition waste in Norway. J Ind Ecol 11:27–39

Bridgwater AV (1986) Refuse composition projections and recycling technology. Resour Conserv 12:159–174

Bruvoll A, Ibenholt K (1997) Future waste generation: forecasts on the basis of a macroeconomic model. Resour Conserv Recycl 19:137–149

Buenrostro O, Bocco G, Vence J (2001) Forecasting generation of urban solid waste in developing countries—a case study in Mexico. J Air Waste Manage Assoc 51:86–93

Chaerul M, Tanaka M, Shekdar AV (2008) A system dynamics approach for hospital waste management. Waste Manag 28(2008):442–449

Chang NB, Lin YT (1997) An analysis of recycling impacts on solid waste generation by time series intervention modeling. Resour Conserv Recycl 19:165–186

Chen HW, Chang N (2000) Prediction analysis of solid waste generation based on grey fuzzy dynamic modeling. Resour Conserv Recycl 29:1–18

Chen M, Giannis A, Wang JY (2012) Application of system dynamics model for municipal solid waste generation and landfill capacity evaluation in Singapore. Macro Theme Rev A Multidisciplinary J Global Macro Trends 1(1): 101–114

Cherian J, Jacob J (2012) Management models of municipal solid waste: a review focusing on socio-economic factors. Int J Econ Finan 4(10):131–139

Chu Z, Wu Y, Zhou A, Huang W-C (2016) Analysis of influence factors on municipal solid waste generation based on the multivariable adjustment. Environ Progr Sustain Energy

Chung S-S, Poon C-S (1998) A comparison of waste management in Guangzhou and Hong Kong. Resour Conserv Recycl 22:203–216

Chung -S-S (2010) Projecting municipal solid waste: the case of Hong Kong SAR. Resour Conserv Recycl 54(11):759–768

Chunsheng G (2009) Relationship between consumption patterns and waste composition, industrial ecology. Royal Institute of Technology, Stockholm

Dangi MB, Pretz CR, Urynowicz MA, Gerow KG, Reddy JM (2011) Municipal solid waste generation in Kathmandu, Nepal. J Environ Manag 92:240–249

Dasgupta D, Debsarkar A, Hazra T, Bala B, Gangopadhyay A, Chatterjee D (2016) Scenario of future e-waste generation and recycle-reuse landfill-based disposal pattern in India: a system dynamics approach. Environ Dev Sustain 18(3)

Daskalopoulos E, Badr O, Probert SD (1998) Municipal solid waste: a prediction methodology for the generation rate and composition in the European Union countries and the United States of America. Resour Conserv Recycl 24:155–166 (1998)

Dennison GJ, Dodd VA, Whelan B (1996) A socio-economic based survey of household waste characteristics in the city of Dublin, Ireland, I. Waste composition. Resour Conserv Recycl 17 (3):227–244

Dyson B, Chang N-B (2005) Forecasting municipal solid waste generation in a fast-growing urban region with system dynamics modelling. Waste Manag 25(7):669–679

Eleyan D, Al-Khatib IA, Garfield J (2013) System dynamics model for hospital waste characterization and generation in developing countries. Waste Manage Res 31(10):986–995

Gautam AK, Kumar S (2005) Strategic planning of recycling options by multi-objective programming in a GIS environment. Clean Technol Environ Policy 7(4):306–316

Gay AE, Beam TG, Mar BW (1993) Cost-effective solid waste characterization methodology. J Environ Eng 119:631–644

Gidarakos E, Havas G, Ntzamilis P (2006) Municipal solid waste composition determination supporting the integrated solid waste management system in the Island of Crete. Waste Manag 26:668–679

Goel S (2008) Municipal solid waste management (MSWM) in India: a critical review. NEERI JESE 50(4):319–328

Gómez G, Meneses M, Ballinas L, Castells F (2009) Seasonal characterization of municipal solid waste (MSW) in the city of Chihuahua, Mexico. Waste Manag 29:2018–2024

Granger CWJ (1989) Forecasting in business and economics, 2nd edn. Academic Press, San Diego

Grazhdani D (2016) Assessing the variables affecting on the rate of solid waste generation and recycling: an empirical analysis in Prespa Park. Waste Manag 48(2016):3–13

Grossman D, Hudson JF, Marks DH (1974) Waste generation models for solid waste collection. ASCE J Environ Eng Div 100:1219–1230

Hibiki A, Shimane T (2006) Empirical study on determinants of household solid waste and the effect of the unit pricing in Japan. In: IEEE proceedings of the annual meeting of Japanese Economic Association, Osaka City University, 21–22 October 2006, Osaka, Japan

Hockett D, Lober DJ, Pilgrim K (1995) Determinants of per capita municipal solid waste generation in the Southeastern United States. J Environ Manage 45(3):205–217

Hoornweg D, Bhada-Tata P (2012) A global review of solid waste management. The World Bank

Huang GH, Baetz BW, Patry GG (2005) Grey fuzzy Integer programming: an application to regional waste management planning under uncertainty. Socio-Econ Plann Sci 29:17–38

Hymans SH (2008) Forecasting and econometric models in the concise encyclopedia of economics. http://www.econlib.org/library/Enc/ForecastingandEconometricModels.html. Accessed 6 Aug 2016

IBRD-World Bank (1999) What a waste: solid waste management in Asia. web.mit.edu/urbanupgrading/urbanenvironment/resources/references/pdfs/WhatAWasteAsia.pdf

Intharathirat R, Abdul Salam P, Kumar S, Untong A (2015) Forecasting of municipal solid waste quantity in a developing country using multivariate grey models. Waste Manag 39(1):3–14

Jalili M, Noori R (2008) Prediction of municipal solid waste generation by use of artificial neural networks: a case study of Mashhad. Int J Environ Res 2:22–33

Jang J-SR (1993) ANFIS: adaptive-network-based fuzzy inference system. IEEE Trans Syst Man Cybern 23(3):665–685

Jena SK, Goel S (2014) E-waste generation in an academic campus: IIT Kharagpur as a case study. Pollution Res 34(2):315–320

Joutz FL (1996) Modeling and forecasting municipal solid waste generation in the US energy supply. J Forecasting 15:477–494

Kansal A (2002) Solid waste management strategies for India. Indian J Environ Prot 22(4):444–448

Karadimas NV, Loumos VG (2008) GIS-based modelling for the estimation of municipal solid waste generation and collection. Waste Manage Res 26:337–346

Karadimas NV, Orsoni A (2006) Municipal solid waste generation modelling based on fuzzy logic. In: Proceedings 20th European conference on modelling and simulation

Karavezyris V, Timpe KP, Marzi R (2002) Application of system dynamics and fuzzy logic to forecasting of municipal solid waste. Math Comput Simul 60:149–158

Katsamaki A, Willems S, Diamadopoulos E (1998) Time series analysis of municipal solid waste generation rates. J Environ Eng. 178–183 10.1061/(ASCE)0733-9372(1998)124:2

Keser S, Duzgun S, Aksoy A (2012) Application of spatial and nonspatial data analysis in determination of the factors that impact municipal solid waste generation rates in Turkey. Waste Manag 32:359–371

Khan S, Farooqi IH (2012) Prioritising municipal solid waste management factors in india using fuzzy analytic hierarchy process. Int J Environ Waste Manag 10(4):423–440

Khan MZA, Burney (1989) Forecasting solid waste composition—an important consideration in resource recovery and recycling. Resour Conserv Recycl 3(1):1–17

Khan MZA, Burney FA (1984) Prediction of solid waste generation rates using socio-econo-climatic factors. Civil Eng Pract Design Eng 3(10):1009–1018

King BF, Murphy RC (1996) Survey to estimate residential solid waste generation. J Environ Eng 122(10):897–901

Kollikkathara N, Feng H, Yu D (2010) A system dynamic modeling approach for evaluating municipal solid waste generation, landfill capacity and related cost management issues. Waste Manag 30:2194–2203

Korhonen P, Kaila J (2015) Waste container weighing data processing to create reliable information of household waste generation. Waste Manag 39(2015):15–25

Kumar JS, Rao DN, Sarcar MMM (2005) Prediction of municipal solid waste generation with artificial neural networks model—a case study. Indian J Environ Prot 25(7):595–600

Kumar JS, Subbaiah KV, Rao PVVP (2011) Prediction of municipal solid waste with RBF network—a case study of Eluru, A.P. India. Int J Innovat Manage Technol 2:238–243

Lebersorger S (2011) Municipal solid waste generation in municipalities: quantifying impacts of household structure, commercial waste and domestic fuel. Waste Management

Li PH, Kwon HH, Sun L, Lall U, Kao JJ (2010) A Modified support vector machine based prediction model on stream-flow at the Shihmen reservoir, Taiwan. Int J Climatol 30(8):1256–1268

Liu G, Yu J (2007) Gray correlation analysis and prediction models of living refuse generation in Shanghai city. Waste Manage 27:345–351

Liu ZF, Xi P, Wang SW, Liu GF (2002) Recycling strategy and a recyclability assessment model based on an artificial neural network. J Mater Process Technol 129:500–506

Ma H, Ho Y-S, Fu H-Z (2011) Solid waste related research in science citation index expanded. Arch Environ Sci 5:89–100

Mashayekhi AN (1993) Transition in New York State solid waste system: a dynamic analysis. Syst Dyn Rev 9:23–48

Matsunaga K, Themelis NJ (2002) Effects of affluence and population density on waste generation and disposal of municipal solid wastes. http://www.seas.columbia.edu/earth/wtert/waste-affluence-paper.pdf

Matsuto T, Tanaka N (1993) Data analysis of daily collection tonnage of residential solid waste in Japan. Waste Manage Res 11(4):333–343

McBean EA, Fortin MHP (1993) A forecast model of refuse tonnage with recapture and uncertainty bounds. Waste Manage Res 11(5):373–385

Mohee R, Mauthoor S, Bundhoo ZM, Somaroo G, Soobhany N, Gunasee S (2015) Current status of solid waste management in small island developing states: a review. Waste Manag 43:539–549. doi:10.1016/j.wasman.2015.06.012

Mwenda A, Kuznetsov D, Mirau S (2014) Time series forecasting of solid waste generation in Arusha City—Tanzania. Math Theory Model 4(8):29–39

Navarro-Esbri J, Diamadopoulos E, Ginestar D (2002) Time series analysis and forecasting techniques for municipal solid waste management. Resour Conserv Recycl 35(3):201–214

Neto JC, Silva MM, Santos SM (2016) A time series model for estimating the generation of lead acid battery scrap. Clean Technol Environ Policy 1–13

Niessen WR, Alsobrook AF (1972) Municipal and industrial refuse: composition and rates. In: Proceedings of national waste processing conference, pp 112–117

Noori R, Abdoli M, Ameri Ghasrodashti A, Jalili Ghazizade M (2009b) Prediction of municipal solid waste generation with combination of support vector machine and principal component analysis: a case study of Mashhad. Environ Progress Sustain Energy 28:249–258

Noori R, Abdoli M, Farokhnia A, Abbasi M (2009a) Results uncertainty of solid waste generation forecasting by hybrid of wavelet transform-ANFIS and wavelet transform-neural network. Expert Syst Appl 36:9991–9999

Noori R, Karbassi A, Sabahi MS (2010) Evaluation of PCA and gamma test techniques on ANN operation for weekly solid waste prediction. J Environ Manage 91:767–771

Ojeda-Benıtez S, Vega CA, Marquez-Montenegro MY (2008) Household solid waste characterization by family socioeconomic profile as unit of analysis. Resour Conserv Recycl 52:992–999

Openshaw S, Openshaw C (1997) Artificial intelligence in geography, 2nd edn. Wiley, Chichester

Oumarou MB, Dauda M, Abdulrahim AT, Abubakar AB (2012) Municipal solid waste generation recovery and recycling: a case study. World J Eng Pure Appl Sci 2(5):143–147

Owusu-Sekyere E, Bagah AD, Dwamena Quansah JY (2015) The urban solid waste management conundrum in Ghana: will it ever end? World Environ 5(2):52–62

Pai R, Rodrigues LLR, Mathew AO, Hebbar S (2014) Impact of urbanization on municipal solid waste management: a system dynamics approach. Int J Renew Energy Environ Eng 2 ISSN 2348-0157

Patel V, Meka S (2013) Forecasting of municipal solid waste generation for medium scale towns located in the state of Gujarat, India. Int J Innovat Res Sci Eng Technol 2:4707–4716

Phuntsho S, Heart S, Shon H, Vigneswaran S, Dulal I, Yangden D, Tenzin UM (2009) Studying municipal solid waste generation and composition in the urban areas of Bhutan, obtained in http://www98.griffith.edu.au/dspace/bitstream/handle/10072/30242/62932_1.pdf?sequence=1. Accessed 31st July 2015

Pires A, Martinho G, Chang N-B (2011) Solid waste management in European countries: a review of systems analysis techniques. J Environ Manage 92(2011):1033–1050

Purcell M, Magette WL (2009) Prediction of household and commercial BMW generation according to socio-economic and other factors for the Dublin region. Waste Manag 29:1237–1250

Raimondi FM, Sella M, Italia F, Martinez A (1997) Urban solid waste generation model via fuzzy-genetic algorithm. In: Proceedings international conference on measurements and modelling in environmental pollution, pp 527–537

Rhyner CR, Green BD (1988) The predictive accuracy of published solid waste generation factors. Waste Manage Res 6(1):329–338

Salam MA, Hossain ML, Das SR, Wahab R, Hossain MK (2012) Generation and assessing the composition of household solid waste in commercial capital city of Bangladesh. Int J Environ Sci Manag Eng Res 1(4):160–171

Salhofer S (2001) Kommunale Entsorgungslogistik: Planung. Berlin. Erich Schmidt, Gestaltung und Bewertung entsorgungslogistischer Systeme für kommunale Abfälle. doi:10.1016/j.wasman.2006.12.011

Shahabi H, Khezri S, Ahmad BB, Zabihi H (2012) Application of artificial neural network in prediction of municipal solid waste generation (case study: Saqqez city in Kurdistan Province). World Appl Sci J 20(2):336–343

Shamshiry E, Mokhtar M, Abdulai A-M, Komoo I, Yahaya N (2014) Combining artificial neural network-genetic algorithm and response surface method to predict waste generation and optimize cost of solid waste collection and transportation process in Langkawi Island, Malaysia. Malaysian J Sci 33(2):118–140

Shamshiry E, Nadi B, Bin Mokhtar M, Komoo I, Hashim HS, Ahya NY (2011) Forecasting generation waste using artificial neural networks. In: Proceedings of the 2011 international conference on artificial intelligence, ICAI 2011

Shi JG, Xu YZ (2006) Estimation and forecasting of concrete debris amount in China. Resour Conserv Recycl 49:147–158

Shu HY, Lu HC, Fan HJ, Chang MC, Chen JC (2006) Prediction from energy content of Taiwan municipal solid waste using multilayer perception neural networks. J Air Waste Manag Assoc 56:852–858

Srivastava AK, Nema AK (2005) Grey modelling of solid waste volumes in developing countries. Proc ICE—Waste Resour Manage 159:145–150

Sudhir Kumar J, Venkata Subbaiah K, Prasada Rao PVV (2011) Prediction of municipal solid waste with rbf net work—a case study of Eluru, A.P., India. Int J Innov Manag Technol 2 (3):238–243

Sudhir V, Srinivasan G, Muraleedharan VR (1997) Planning for sustainable solid waste in urban India. Syst Dyn Rev 13:223–246

Sufian MA, Bala BK (2007) Modeling of urban solid waste management system: the case of Dhaka city. Waste Manag 27:858–868

Thanh NP, Matsui Y, Fujiwara T (2010) Household solid waste generation and characteristic in a Mekong Delta city Vietnam. J Environ Manag 91(2010):2307–2321

Tiwari MK, Bajpai S, Dewangan UK (2012) Prediction of industrial solid waste with ANFIS model and its comparison with ANN Model—a case study of Durg-Bhilai Twin City India. Int J Eng Innov Technol (IJEIT) 2:192–201

USEPA (1997) Appendix H: methodology to calculate waste generation based on previous years. In: Measuring recycling: a guide for state and local governments: adjusting waste generation. USEPA 03-27-2003

Vesilind PA, Worrell W, Reinhart D (2002) Solid waste engineering. Thomson Learning Inc., Singapore

Viswanathan C (2006) Domestic solid waste management in South Asia. http://www.faculty.ait.ac.th/visu/ (2006)

Vivekananda B, Nema AK (2014) Forecasting of solid waste quantity and composition: a multilinear regression and system dynamics approach. Int J Environ Waste Manag 13(2):179–198

Wang H, Nie Y (2001) Municipal Solid waste characteristics and management in China. J Air Waste Manag Assoc 51(2):250–263

Wei Y, Xue Y, Yin J, Ni W (2013) Prediction of municipal solid waste generation in China by multiple linear regression method. Int J Comput Appl 35(3):136–140

Wieland D, Wotawa F, Wotawa G (2002) From neural networks to qualitative models in environmental engineering. Comput Aided Civil Infrastruct Eng 17:104

Xu L, Gao P, Cui S, Liu C (2013) A hybrid procedure for MSW generation forecasting at multiple time scales in Xiamen City China. Waste Manage 33:1324–1331

Xu L, Lin T, Xu Y, Xiao L, Ye Z, Cui S (2016) Path analysis of factors influencing household solid waste generation: a case study of Xiamen Island, China. J Mater Cycles Waste Manage 18(2):377–384

Yang L, Chen Z, Liu T, Gong Z, Yu Y, Wang J (2013) Global trends of solid waste research from 1997 to 2011 by using bibliometric analysis. Scientometrics 96:133–146

Ying L, Weiran L, Jingyi L (2011) Forecast the output of municipal solid waste in Beijing Satellite Towns by combination models. In: 2011 International conference electric technology and civil engineering (ICETCE). IEEE, pp 1269–1272

Yost P, Halstead J (1996) A methodology for quantifying the volume of construction waste. Waste Manage Res 14:453–461

Younes MK, Nopiah ZM, Ahmad Basri NE, Basri H, Abushammala Mohammed FM, Maulud KNA (2015) Solid waste forecasting using modified ANFIS modeling. J Air Waste Manag Assoc 65(10):1229–1238. doi:10.1080/10962247.2015.1075919

Yousuf TB, Rahman M (2007) Monitoring quantity and characteristics of municipal solid waste in Dhaka City. Environ Monit Assess 135:3–11

Zeng Y, Trauth MK (2005) Internet-based fuzzy multicriteria decision support system for planning integrated solid waste management. J Environ Inform 6(1):1–15

Shale Gas: Hydrofracking, its Effects and Possible Remediation

Waheed Gbenga Akande

Abstract Unconventional petroleum and natural gases such as shale oils and shale gases have been considered in recent times as the alternative energy sources to augment global conventional petroleum reserve to keep pace with global energy demand considering enormous reserves of these resources all over the world. The United States of America is the pioneer in unconventional petroleum exploration, development, and production, and the lessons learnt in the US are now being transferred to other parts of the world to develop their unconventional petroleum resources. Horizontal drilling and hydraulic fracturing (hydrofracking or hydrofrac) are the emerging technologies to exploit gas shale resources. Hydraulic fracturing operations for shale gas production have certain effects such as surface water and groundwater contaminations, undue pressure on public water supply, greenhouse gas emissions, and seismic events on the local environments, man, ecosystems, and groundwater systems; and the severity of these impacts depends on chemical composition of shales/shale gases, wellbore integrity, chemicals used in fracking fluids and management of various stages of shale gas development and production. It is suggested as a remediation measure that the environmental impact assessment (EIA) should be carried out by interested companies in shale gas business and their reports submitted at an early exploration stage to the host government or relevant agencies. Environmental-related legislation and regulation as well as safety guidelines for drilling and production of shale gas such as disclosure of chemicals in fracking fluids should be enforced. Intensive and collaborative researches should be supported by various energy stakeholders such as governments, academia, shale gas prospectors, and non-governmental organizations (NGOs) to understand the chemical constituents of shale deposits and shale gases, monitor the impacts of shale gas extraction on the groundwater system, and determine possible connection between hydrofracking and seismic events. It is concluded that with this intensified research into shale gas exploration, development and production in mind, better technologies which are more environmentally friendly to exploit gas shale assets will soon emerge.

W.G. Akande (✉)
Department of Geology, Federal University of Technology, Minna, Nigeria
e-mail: geowaheed2008@yahoo.com; waheed.akande@futminna.edu.ng

D. Sengupta and S. Agrahari (eds.), *Modelling Trends in Solid and Hazardous Waste Management*, DOI 10.1007/978-981-10-2410-8_4

Keywords Shale deposits · Shale gas · Hydrofracking · Fracking fluids · Wastewater · Fugitive gases · Contamination · Environmental impact assessment (EIA)

Introduction

The current trend in decline in global petroleum reserve and inability of the conventional petroleum and natural gases to keep pace with global energy demands has necessitated looking for other alternative energy sources. Recently, this has made oil and gas industries, academia, researchers, and governments across the globe to intensify their efforts in real-time investments in unconventional petroleum sources such as shale oils and shale gases. These efforts are justified considering the proved enormous reserves of unconventional petroleum in many parts of the world such as in the United States of America, European countries, and even now in developing nations. The United States of America has been in forefront in the business of unconventional petroleum with the giant strides to develop the resource beginning as far back as the late 1970s (Wang and Krupnick 2013). Shale gas production accounted for only 1.4 % (approx. 7.6 billion cubic meters, 7.6 bcm) of total US natural gas production in 1990 (EIA 2010), 1.6 % in 2000 (Wang and Krupnick 2013), 4.1 % by 2005 (Wang and Krupnick 2013), 14.3 % (93 bcm) in 2009 (EIA 2010b), and skyrocketed to an astonishing 23.1 % by 2010 (Wang and Krupnick 2013). This remarkable growth of shale gas development and production in the United States remains an impetus for increasing interest in exploring shale resources in other parts of the globe. A number of countries, including China, Mexico, Argentina (Gonzalez 2012; Orihuela 2012), Poland, India, and Australia (IEA 2012) are currently considering or are in the process of developing their own shale gas resources (Wang and Krupnick 2013).

Shale gas has been defined as a natural gas that is trapped within fractures and pore spaces within fine-grained sedimentary shale rocks (Llewellyn 2009). As opposed to conventional petroleum where we have independent petroleum sources (source rocks) and reservoirs, and the petroleum expulsion and migration take place from the former into the latter; hydrocarbons mainly natural gas of predominantly methane composition are only stored in the matrix of unconventional petroleum sources (shales).

Shale gas is one of the unconventional petroleum sources (others being coal bed methane, CBM, shale oil, tar sands, gas hydrates, tight gas, etc.) that currently attract attention to increase global petroleum reserve base. Shale gas extraction and its operational requirements are to some extent slightly different from those of other unconventional petroleum sources. The advances in technologies have revolutionized the way shale gas extraction is now being carried out worldwide. Hydraulic fracturing, hydrofracking or hydrofrac, has become an emerging technology that is now being employed for shale gas extraction. This chapter looks into the aspect of shale gas extraction using hydrofracking, its effects on man, local environments, and the ecosystem in the shale gas fields arising from both natural and mechanical defects during methane gas extraction, possible environmental implications and remediations.

Principle of Hydrofracking and Shale Gas Extraction

The principle of hydrofracking is based on creation of macrofractures in shale source rocks to connect pores, fissures, and microfractures thereby making migration and flow of shale gas to the wellbore possible. Two complementary drilling techniques have been identified and are now in use for shale gas extraction, and these are horizontal drilling and hydraulic fracturing (Fig. 1). These techniques are in addition to traditional vertical well drilling which the operation normally starts with. Figure 2 shows different components of a modern shale gas well.

According to the Tyndall Centre (2011), these drilling techniques are used in combination with one another to extract shale gas and they are briefly explained below:

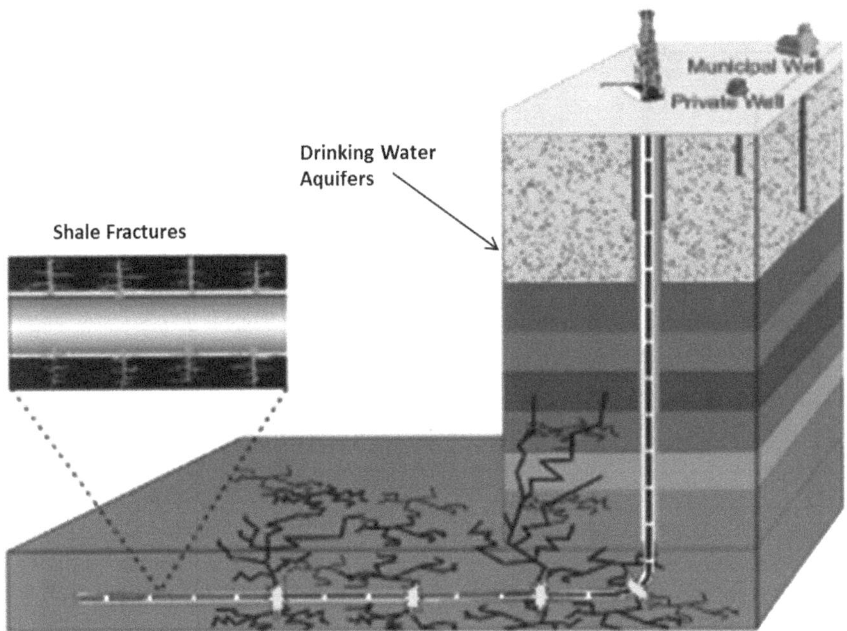

Fig. 1 Diagram of horizontal fracking (*Source* EPA)

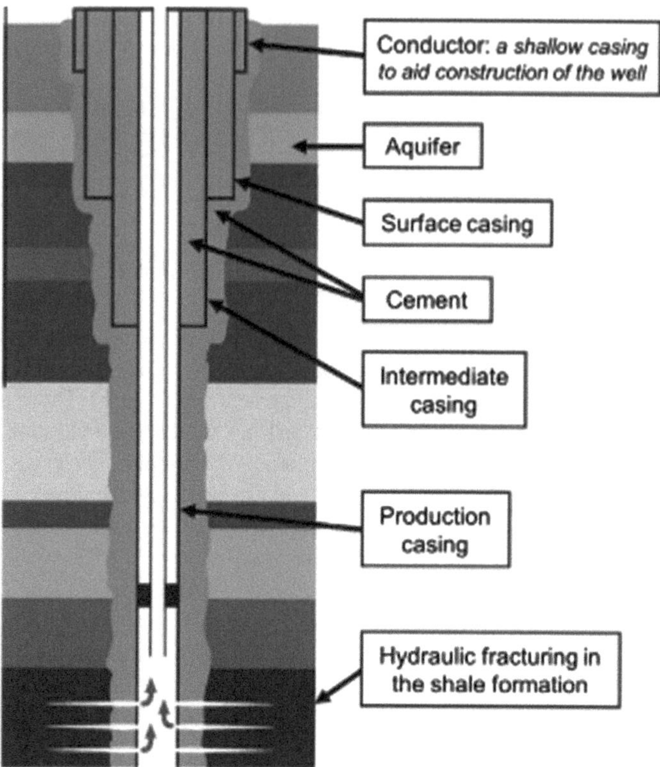

Fig. 2 Typical modern design of shale gas well (based on Cuadrilla design 2012)

> Horizontal drilling is used to provide greater access to the gas trapped deep in the producing formation. At the desired depth, the drill bit is turned to bore a well that stretches through the reservoir horizontally, exposing the well to more of the producing shale;
>
> Hydraulic fracturing is where fluid (water, sand, and other substances) are pumped into the well at pressure to create and increase fractures in the rock. These fractures start at the injection well and can extend a few hundred metres into the reservoir rock. A material such as sand holds the fractures open, allowing hydrocarbons to flow into the reservoir rock. Between 15 and 80 % of the injected fluids are recovered at the surface. Fluid that returns to the surface is captured, treated and disposed of and gas that flows to the surface is captured and used for electricity generation or is put into the mains supply. It is also possible to 'frack' a well several times in its lifetime to increase yield.

In shale gas extraction, the injected fracturing fluid which is usually water-based with small amounts of silica (sand) or similar particulate matter are normally introduced into the source bed (target formation) so as to prop open the fractures. The fractures may be in order of a few micrometers in width and usually limited in length to a few tens of meters. Upon creation of an artificial fracture, individual molecules of shale gas that are far away from the well can find their way to the fractures, and once there, can migrate quickly through the fractures to the wellbore

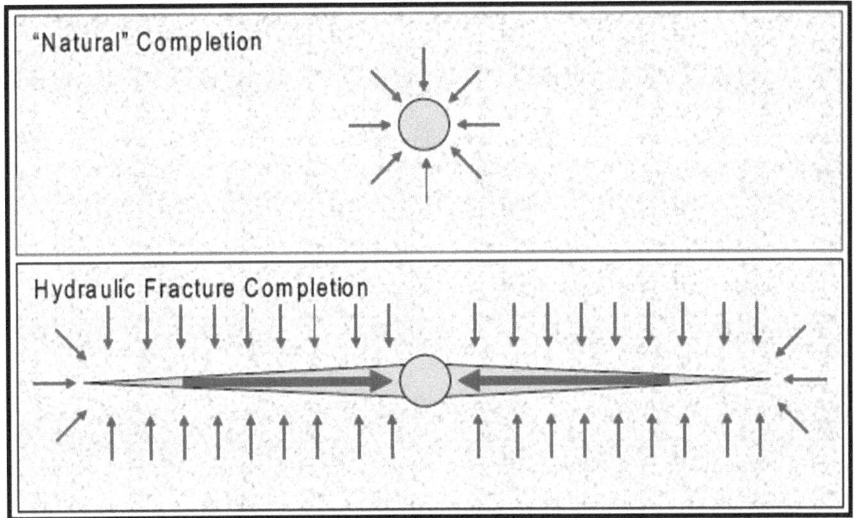

Fig. 3 Illustration of a fractured and a nonfractured well. The *top portion* of the figure shows "traditional" nonfractured well, where the *red arrows* represent the flow of fluid to the circle which represents the well. The *bottom part* depicts that by creating an artificial fracture, individual molecules that are a long distance from the well can find their way to the fracture, and ultimately travel quickly through the fracture to the well (API Guidance Document HF1, 2009)

(Fig. 3). Further details of hydraulic fracturing operations can be found in API Guidance Document HF1 (2009).

Effects of Hydrofracking Process

The resulting effects of the operations involved in shale gas extraction on man and ecosystems as well as groundwater systems depend on certain factors. The most important of these is the chemical composition of the source rocks (shales). Source composition is also dependent on the depositional environments ranging from continental through lacustrine to marine environments. For instance, marine shales are likely to be richer in sulfur than lacustrine shales; and this will have a significant effect on the fracturing environment and ecosystems. Thus, the compositions of shales and resulting shale gas could vary widely. Hydraulic fracturing fluid compositions and the manner in which the resulting wastewater is managed and disposed are other key factors. The overall management of all the phases involved in shale gas extraction, development, and production is also critical. These factors may cause hydrofracking operations to pose serious problems to man, ecosystems, and groundwater systems.

Although hydraulic fracturing has been described as an established technology (The Royal Society and the Royal Academy of Engineering 2012), concerns have been expressed that the extraction of unconventional gas through this process may be detrimental to both the environment and local communities (Tyndall Centre 2011). A number of possible environmental impacts have been identified and these are discussed below.

Environmental Impacts Related to Water Used in Shale Gas Operations

It has been reported that shale gas development requires large volumes of water (e.g., Tyndall Centre 2011). It follows that in the regions where the water resource is not readily adequate shale gas exploration may put the resource under pressure. As the technology for shale gas development is being exported to the developing countries where in most cases water supply is grossly inadequate, the activity may further reduce the volumes of water available to people for their various daily needs. Access to water sources is likely to become more of a constraint for operators in arid regions in particular facing growing depletion of water resources, and in areas where water flows and availability follow seasonal variations (Stark et al. 2012).

Environmental impacts associated with water and drilling operations include cross-contamination of underground aquifers due to poor borehole construction when the integrity of wellbore casing is compromised; pollution from an unexpected release of gas or fracturing fluids (especially surfactants, biocides, and mineral acid for acidizing process) into other parts of the water environment can adversely affect animals and ecosystems due to their toxic nature; and surface contamination from the uncontrolled disposal of liquid (wastewater) or solid waste containing potentially harmful substances; and abstraction of uncontrolled volumes of water which could lead to an unacceptable impact on the water environment and its ecosystems.

According to Stamford and Azapagic (2014), although the impacts of fracking fluids on groundwater system is still a matter of debate in the literature, the impacts from fracking fluid are centered on the potential contamination of groundwater owing to accidents and/or malpractice. This could involve contamination with fracking fluid components or naturally occurring substances that have been mobilized by the extraction process, such as heavy metals (Stamford and Azapagic 2014).

Environmental Impacts Related to Induced Seismic Events

The fracturing operations for shale gas exploration have also been suspected to be related to the occurrences of low magnitude seismic events such as earthquakes. A good example in this regard was a recent report produced for the Department of

Energy and Climate Change (DECC) in 2012 which concluded that the seismic events recorded at Preese Hall in Lancashire were directly connected to the fracturing operations. Stamford and Azapagic (2014) also reported that in the UK, low-intensity earthquakes (measuring 2.3 and 1.5 on the Richter scale) were observed in April 2011 due to fracking in North West England which led the government to suspend shale gas extraction nationally from May 2011 to December 2012, because there was an impression that the seismic event was attributable to hydraulic fracturing. Though real data evidences are to yet be provided by the researchers or seismologists in this field, presupposed induced seismic events could cause damage to facilities and even death depending on the severity of the events.

Environmental Impacts Related to Greenhouse Gas Emissions

Shale gas exploitation and production activities have a potential for increased greenhouse gas emissions from fugitive releases. The role of methane gas in greenhouse and global warming phenomena is often underrated compared to carbon dioxide. However, methane could be a more potent greenhouse gas than carbon dioxide. It has been reported that fugitive releases of methane during shale gas operations is higher than those of conventional gas but less than from coal, and this observation awaits empirical data to substantiate it (EU Report for European Commission 2012). Related to this is the impact of the shale gas compositions. Shale gases containing appreciable amounts of sulfur or hydrogen sulfide and nitrogen can be oxidized into respective acids thereby causing havocs to facilities especially if they are designed by not taking this factor into consideration. In addition, if these gases are dissolved in or exsolved into the groundwater, they are capable of contaminating the water or increasing the acidity of the aquifer systems. Finally, if they find their way into the Earth's atmosphere they play a role of greenhouse gases in the form of oxides of sulfur (e.g., SO_2) and nitrogen (e.g., NO_2).

Remediations

In order to nip the environmental impacts of shale gas exploitation, development, and production in the bud, the environmental impact assessment (EIA) of the shale gas development on sites is critical. It is the role of the host government and its agencies, and other relevant stakeholders in environmental-related matters to request for the EIA Reports from the companies licensed with shale gas plays at an early exploration stage before actual development and production commence. These government agencies should mandate the shale gas operators to disclose the chemicals of the fracturing fluids. This is already a common practice in the United States of America and some European countries (e.g., Poland now makes fracking fluid chemicals disclosure mandatory). Although currently not obligated, operators in South Africa are committed to voluntarily disclose the chemicals used in hydrofracking and it is expected that in the near future, many other countries especially China and Argentina shall embrace some level of disclosure (Stark et al. 2012).

An adequate knowledge of the chemical composition of the shale deposits is very paramount to prediction of the compositionally related impacts of shale gas exploitation on man, ecosystems, and groundwater systems and this has to be established. The shale gas fields should be put into the regional geology context in order to understand the depositional environment of the shale source beds. This should also be approached through chemical analysis of the shale formations. The analytical results from the chemical analysis also have certain implications for shale gas development such as shale gas well design and completion.

With respect to the disposal of excess fracturing fluid and residual fluids produced during shale gas production, wastewater ponds should be discouraged while closed metal tanks should be made available for temporary storage of wastewater before it is treated for disposal or reuse. This is because the wastewater ponds and open pits can lead to local environmental damage should the pits overflow in the event of heavy rainfall (POSTNOTE 374 April 2011). Closed metal tanks are now made compulsory for temporary storage of wastewater in the UK (The Royal Society and the Royal Academy of Engineering 2012). It is currently agreed that shale gas extraction process requires huge quantities of water. In order to lessen this burden on public water supply, researchers should continue to advance the technology (hydraulic fracturing) for shale gas exploitation to reduce its water requirement. Appropriate legislation and regulation should be put in place to deter fugitive emissions which are potential greenhouse gases during drilling and production of shale gases. Finally, the governments, academia, shale gas prospectors, energy operators, and non-governmental organizations (NGOs) should support and fund researches on the possible linkage between shale gas exploitation activities and seismic events.

Conclusions

This chapter appraised hydraulic fracturing as an emerging technique for shale gas exploitation and development and the effects of the overall processes on the local inhabitants, man, ecosystems as well as surface and groundwater systems. The identified effects of hydrofracking include pressure on public water supply and surface water, groundwater pollution or contaminations from wastewater and residual fracking fluids, release of fugitive gases which are potential sources of greenhouse gases into the Earth's atmosphere, and seismic events.

It is suggested as a remediation measure that the environmental impact assessment (EIA) should be carried out by the companies licensed to develop gas shale deposits and their reports submitted at an early exploration stage to the host government or relevant agencies, and this has to be concluded before actual development and production of shale gas begins. Legislation and regulation and safety guidelines for drilling and production of shale gas such as chemical in fracking fluids disclosure should be enforced. Researches should be supported by various stakeholders such as governments, academia, shale gas prospectors, energy operators, and non-governmental organizations (NGOs) to understand the chemical constituents of shale deposits and shale gases, monitor the influence of shale gas exploitation on the groundwater system, and determine possible connection between hydrofracturing and seismic events. It is believed that with dogged researches to review and advance the current hydraulic fracturing technique, better technologies which are more environmentally friendly to exploit gas shale assets will soon emerge.

References

API Guidance Document HF1 (2009) Hydraulic fracturing operations—well construction and integrity guidelines, 1st edn. Publication of American Petroleum Institute, Oct 2009

Climate impact of potential shale gas production in the EU Report for European Commission (2012) DG CLIMA AEA/R/ED57412

Cuadrilla Resources Ltd. (2012) Wellbore integrity—Cuadrilla land based wells. Cuadrilla Resources Ltd, Lichfield

Energy Information Administration (2010). Annual Energy Outlook 2011: early release overview, Published December 16, 2010

EPA: The United States Environmental Protection Agency

Gonzalez P (2012) YPF said to expect approval to double argentine gas prices. Bloomberg. Accessed 25 March 2015

IEA (2012) International energy agency. Gas: Medium-Term Market Report 2012, p 61

Llewellyn L (2012) Shale gas and coal-bed methane (unconventional gas). A publication of National Assembly of Wales, Paper number: 12/041

Orihuela R (2012) Argentina seizes oil producer YPF, as Repsol Gets Ousted. Bloomberg. Accessed 25 March 2015

POSTNOTE 374 April 2011 Unconventional Gas

Stamford L, Azapagic A (2014) Life cycle environmental impacts of UK shale gas. Appl Energy 134:506–518

Stark M, Allingham R, Calder J, Lennartz-Walker T, Wai K, Thompson P, Zhao S (2012) Water and shale gas development leveraging the US experience in new shale developments. Accenture—High Performance Delivered

The Royal Society and the Royal Academy of Engineering (2012) Shale gas extraction in the UK: a review of hydraulic fracturing

Tyndall Centre (2011) Shale gas: a provisional assessment of climate change and environmental impacts

Wang Z, Krupnick A (2013) A retrospective review of shale gas development in the United States: what led to the boom? resource for the future: Washington, DC20036

Characterization and Monitoring of Solid Waste Disposal Sites Using Geophysical Methods: Current Applications and Novel Trends

Pantelis Soupios and Dimitrios Ntarlagiannis

Abstract Landfilling remains the most attractive waste management method for solid waste. Although not the most efficient and environmental-friendly option, landfills offer a cost-efficient solution compared to other alternatives. For any landfill to be successful site selection, construction, operation, and post-closure monitoring is critical. Synergistic use of geophysical methods and traditional point sampling (e.g., borehole sampling) allows for high resolution characterization and monitoring of landfills during all stages of operation; from guided site selection, to construction integrity and waste characterization, to leachate recirculation and leak monitoring. Geophysical methods offer advantages, such as high temporal and spatial resolution, non (or minimally) invasive and cost-efficient operation, rendering them a very powerful tool for characterization, and long-term monitoring of waste disposal sites. Since geophysical methods involve the indirect imaging of the subsurface cautious implementation, including direct sampling, is needed for successful application. Multiple geophysical methods have been shown to be suitable for landfill characterization and monitoring. Electrical (resistivity, induced polarization, and self potential) and electromagnetic (transient electromagnetic methods, ground penetrating) are the common geophysical methods employed in waste management operations due to the increased conductivity of waste and leachate. Seismic methodologies can also be used to describe subsurface geology and possible waste horizons. In certain cases, magnetic measurements can also be used for the monitoring and characterization of landfills. Typically, geophysical methods are used to:

- spatially delineate landfills and define landfill geometry,
- monitor and characterize the spatial distribution of moisture, gas content, and leachate inside landfills,

P. Soupios (✉)
Department of Environmental and Natural Resources Engineering, Technological Educational Institute of Crete, 3 Romanou, Chalepa, 73133 Chania, Crete, Greece
e-mail: soupios@staff.teicrete.gr

D. Ntarlagiannis
Department of Earth & Environmental Sciences, Rutgers University, 101 Warren Street, Newark, NJ 07102, USA

D. Sengupta and S. Agrahari (eds.), *Modelling Trends in Solid and Hazardous Waste Management*, DOI 10.1007/978-981-10-2410-8_5

- identify classes of buried waste based on material composition,
- monitor the integrity of the liner, and
- identify and monitor leachate leaks, and the associated contamination plumes.

With this chapter we aim to introduce common geophysical methods and provide examples for application in landfills. For the geophysical methods of interest the basics principles, along with up to date references are provided, and the advantages and limitations for waste management operations are discussed.

Introduction

Integrated waste management (IWM) can be defined as the selection and application of suitable techniques, technologies, and management programs to achieve specific waste management objectives and goals (Tchobanoglou and Kreith 2002). Although the specific objectives and approaches of IWM might differ, the primary target is to reduce the amount of solid waste end in landfills through waste reduction, reuse, recycling, and waste to energy programs. Solid waste management is a complicated process that starts with the proper site selection, and ends with post-closure monitoring that extents long after (decades) the closure of the landfill. Geophysical methods can be used in all stages of landfill operation to provide high resolution characterization and or monitoring. In most cases geophysical methods are used to provide information on the geometrical characteristics of the repository, and indirect information on the physiochemical properties of the infill (solid and leachate) (Tsourlos et al. 2014).

Site Selection and Landfill Construction

Geophysical surveys have been proven to be successful in the landfill site selection stage (Benson 1988). Geophysical characterization can identify areas that are not suitable for landfill construction due to geology (e.g., faults, fracture zones), former mining operations, karst, and high permeable formations. During site construction geophysical methods can be used to confirm the integrity of the containment basin and impermeable liners.

Leachate Monitoring

The most commonly used solid waste classification system separates the waste in three categories (Table 1) with the majority of the landfills designed for municipal solid waste (MSW). MSW is defined as the waste that comes from residential,

Table 1 Classification of landfills based on waste (from Tchobanoglou and Kreith 2002)

Class	Type of waste
I	Hazardous waste
II	Designated waste
III	Municipal solid waste (MSW)

commercial, institutional, and some industrial sources, but exclude hazardous materials. Although MSW appears to be free of dangerous materials, in reality very often contains toxic, and other unsafe, substances. In some instances, the discarded objects might be safe in the original form, but after the deposition in landfills they can become toxic, or release toxic byproducts, due to decomposition or degradation processes. The toxic chemicals can then be mixed with the available water and form the landfill leachate. Landfill leachate can pose a significant environmental and ecological risk, and in some cases even threaten human health. The most common leachate contamination incidents are the result of [a] poorly constructed and/or managed landfill, (Benson et al. 1988; Robinson and Gronow 1995; Hix 1998; Statom et al. 2004), [b] illegal landfills and dumping sites, [c] lack of regulation and enforcement policies (Piratoba Morales and Fenzi 2000; Ntarlagiannis et al. 2016).

To minimize the contamination risk in landfills the implementation of a comprehensive monitoring system is of paramount importance. A properly operated monitoring system can capture any leaks early in time, and prevent extensive damage. In older, typically unlined, landfills the use of a monitoring system is even more important. We should emphasize the landfill monitoring system should stay in place for decades after landfill closure, until the degradation and decomposition processes cease.

Landfill Dynamic Monitoring

During landfill operation geophysical methods offer the spatial and temporal resolution needed to provide information on the main elements of landfills. For example information on boundaries and nature of waste, the depth/thickness and dip of the layers of refuse and sealing materials, the integrity and shape of the capping zones or separating walls and basal floor slopes can be retrieved; furthermore continuous information on leachate, moisture and gas content can be collected, allowing the more efficient operation of bioreactor landfills (Carpenter et al. 1991; Aristodemou and Thomas-Betts 2000; Meju 2000a, b).

Illegal and Abandoned Landfills

Waste disposal sites, especially older ones, are not always properly constructed and monitored; in many cases there are illegal landfills in unknown locations and with unknown characteristics. Geophysical methods can be utilized in such cases to identify and characterize such waste disposal sites. In case of old, abandoned landfills historical information regarding the waste and construction might be missing; geophysical methods can be used to map the waste, and identify leachate pooling and leaks; in some cases, the waste class might be identified. In case of illegal landfills, geophysical methods can first identify the location, and then try to provide information on subsurface characteristics. We should always keep in mind that geophysical methods provide indirect information, so direct sampling could significantly enhance interpretation.

The complexity of the phenomena linked to waste disposal sites necessitates the synergistic use of a variety of characterization and monitoring methods, with geophysical tools playing an important role. Established direct sampling (e.g., borehole sampling) are required for providing detailed and accurate information on waste status and processes. Such information is typically sparse (in space and time); geophysical methods can be then utilized to provide a complete subsurface image of landfill status by constraining the spatially and temporally extensive—even continuous—geophysical data with the detailed direct sampling ones.

Geophysical Methods

Electrical and electromagnetic methods are the most popular geophysical methods used in waste management operations due to their sensitivity in conductivity contrasts; such contrasts are very common in waste deposition sites due to mixed waste, and leachate formation (Meju 2000b; Tsourlos et al. 2014). Common electrical methods (geoelectrical) employed in landfill studies are the direct current (DC) resistivity, induced polarization, (IP) and self potential (SP) (e.g., Naudet 2003a; Rubin and Hubbard 2005; Arora et al. 2007; Soupios et al. 2007a, b, 2008; Grellier et al. 2008; Reynolds 2011; Gazoty et al. 2012; Kemna et al. 2012; Revil et al. 2012b; Belghazal et al. 2013; Tsourlos et al. 2014; Genelle et al. 2014; Vargemezis et al. 2015; Wang et al. 2015; Çınar et al. 2016; Konstantaki 2016). Common EM methods include transient electromagnetic (TEM), radio- or audio-frequency magnetotelluric (RMT/AMT) (Mack and Maus 1986; Tezkan et al. 1996; Meju 2000b; Belghazal et al. 2013; Belmonte-jiménez et al. 2014). Furthermore ground penetrating radar (GPR) and seismic methods can be used to characterize subsurface boundaries and other structural features, including geological ones (Pellerin 2002; Porsani et al. 2004; Rubin and Hubbard 2005; Shemang et al. 2011; Wang et al. 2015). Finally, magnetic surveys can be utilized in cases where ferromagnetic objects are buried (e.g., drums) (Meju 2000b; Prezzi et al. 2005; Huliselan et al. 2010; Belghazal et al. 2013; Almadani et al. 2015) while gravimetric ones can be used when density contrasts are present (Whiteley and Jewell 1992).

In the next sections we will briefly describe the principles of common geophysical methods, provide references for in depth study, and discuss case studies of applications in environmental waste management.

Surface Applications

Electrical Resistivity Imaging (ERI)

Electrical resistivity imaging (ERI) aims at determining the spatial distribution of resistivity ρ in the subsurface, typically with the use of four electrode measurements (Rubin and Hubbard 2005). One pair of electrodes is used for current injection and a pair of electrodes is used to measure the potential difference (Reynolds 2011). Multichannel ERI systems allow for the simultaneous measurement of multiple pairs of potential electrodes; most modern instruments allow the automated acquisition of a sequence of measurements, permitting the creation of 2D, and even 3D, images of the subsurface apparent resistivity (Rubin and Hubbard 2005; Çınar et al. 2016; Konstantaki 2016; Ntarlagiannis et al. 2016). Inverse methods can be used to determine the true subsurface resistivity image (Rubin and Hubbard 2005).

A variety of standard electrode configurations have been developed over the years that offer different survey characteristics and can be suited for different applications; in addition, custom made sequences can be utilized that address the specific objectives of the project (Rubin and Hubbard 2005; Reynolds 2011; Tsourlos et al. 2014). ERI measurements are acquired at various electrode spacing and positions to provide information at various lateral and vertical locations of the study area. Typical applications of the ERI methods involve characterization and monitoring for saltwater and contaminant plumes (Mack and Maus 1986; Slater et al. 2000; Slater 2007; Heenan et al. 2015; Ntarlagiannis et al. 2016), mapping and characterization of buried waste, characterization of engineered structures (e.g., landfill boundaries) (Tsourlos et al. 2014), geological characterization (Robinson et al. 2015a, b), and leak detection and monitoring (Johnson and Wellman 2015).

Time Domain-Induced Polarization (IP)

The induced polarization (IP) method is a natural extension of the resistivity methods whereas not only the resistive, but also the capacitive properties of the earth are measured (Rubin and Hubbard 2005; Reynolds 2011). In certain cases, IP surveys offer additional information about the subsurface, while in general it is more time consuming than ERI; it should be highlighted that during IP surveys ERI data are inherently collected.

Field application of the IP method is similar to the ERI method, where four electrodes are used (two for current injection, and two for potential measurement); additional care should be taken to utilize electrode configurations that provide high S/N ratio, and maintain good contact with the ground (e.g., Mwakanyamale et al. 2012). In addition to the voltage difference measured during ERI surveys, in an IP survey the voltage decay with time, after current injection is stopped, is measured. The recorded gradual voltage decrease is a complex function of charge polarization at the interfaces (e.g., fluid-grain) and charge conduction within the fluid and along the grain (Rubin and Hubbard 2005). The IP method has its origins from ore prospecting, specifically for disseminated metallic minerals (Reynolds 2011; Kemna et al. 2012). Advances in instrumentation, along with better understanding of the underlying processes, led to the resurrection of the IP method in the past couple of decades; IP is now more routinely used in environmental, and other near surface geophysical applications, due to the unique sensitivity in interfacial processes (Rubin and Hubbard 2005; Kemna et al. 2012; Revil et al. 2012b; Abdulrahman et al. 2016; Günther and Martin 2016; Ntarlagiannis et al. 2016).

Self Potential (SP)

Self potential (SP), also known as spontaneous potential, is a passive geophysical method that measures naturally occurring electrical field in earth. SP involves only the use of two, or more, nonpolarizable electrodes (Petiau 2000; Linde et al. 2011); no active signal source is used, as implied by the term 'passive'. SP signals can be caused by multiple processes such as electrokinetic mechanisms (Revil et al. 2003, 2012b), temperature gradients (Reynolds 2011; Revil et al. 2012a), and electrochemical mechanisms (Naudet 2003a; Revil 2003; Reynolds 2011); there is still some uncertainty on the exact physical processes associated with SP signal generation (Reynolds 2011).

Common applications of the SP method involve groundwater movement monitoring, cave detection, sinkhole mapping, contaminant delineation, and leak/seepage detection (e.g., dams) (Rozycki et al. 2006; Suski et al. 2006; Arora et al. 2007). Recently the use of the SP method has been suggested for monitoring microbial processes in the subsurface (Naudet 2003a; Arora et al. 2007; Revil et al. 2010); the proposed model is analogous to the classic geobattery model (Sato and Mooney 1960), but the signal generating processes are biotically driven. Limitations of the SP method are sensitivity to cultural geophysical noise, multiple signal sources, use of specialized electrodes, and difficulties with data processing and interpretation (Reynolds 2011).

Refraction Seismic (RrS)

Seismic refraction is a geophysical method that allows subsurface reconstruction based on the travel properties of P- or S-waves (ASTM 2006; Reynolds 2011). Seismic P- and S-waves generated typically on the surface, then propagate through the soil and rock; a seismograph then is used, along with specific sensors (termed geophones) at known distances from the source, to record the P and S waves. Depths to different layers, and P and S velocities can then be calculated based on recorded arrival times. In general, the refraction method assumes that velocity of the layers increases with depth, thickness of the layers is adequate (depends on survey parameters), and the velocity contrast between layers is sufficient. Common applications of the method include depth to bedrock, geological strata thickness, and subsurface structure characterization. Additionally, several geotechnical parameters and in situ elastic moduli (i.e., bulk and shear modulus, poisson ratio, etc.) can be estimated from RrS (Doll et al. 1996; Soupios et al. 2007a; Almadani et al. 2015; Wijesekara et al. 2015; Benson and Yuhr 2016; Valois et al. 2016).

RrS measurements are sensitive to acoustic noise and vibrations. In cases of low velocity layers which violate the basic assumption of conventional refraction seismic, the refraction tomography method should be applied. For higher resolution, the seismic reflection method can be applied. The primary application of seismic reflection method is the accurate determination of depth and thickness of geologic strata in complex structural environment.

Multichannel Analysis of Surface Waves (MASW) and Spectral Analysis of Surface Waves (SASW)

The MASW and SASW are relatively new in situ seismic methods for determining shear wave velocity profiles. The basis of the methods are the dispersive characteristic of Rayleigh waves when traveling through a layered medium. The Rayleigh wave velocity is determined by the material properties (primarily shear wave velocity, but also compression wave velocity and material density) of the subsurface to a depth of approximately 1–2 wavelengths. Longer wavelengths penetrate deeper and their velocity is affected by the material properties at greater depth. The MASW/SASW methods have significant advantages. The near surface (top 10 m) resolution is typically greater than with other methods. Testing is performed at the ground surface, allowing for a less costly measurement than those carried out with the conventional seismic methods (refraction and reflection). The MASW/SASW testing can be used to obtain Vs profiles for earthquake site response of waste disposal sites and determine soil and rock elastic properties (Greenwood et al. 2015; Yin et al. 2015; Anbazhagan et al. 2016; Gouveia et al. 2016; Ramaiah et al. 2016).

Horizontal to Vertical Spectral Ratio (HVSR)

The usefulness of microtremor (HVSR) measurements as a geophysical tool has been presented by Delgado (Delgado et al. 2000b). The method relies on the relationship between [a] the main resonance frequency (f) of a given soil as obtained from the HVSR of microtremors, [b] its thickness (z) as estimated from other geophysical methods (e.g., ERI, RrS) and [c] the average shear velocity (Vs) according to the following equation:

$$f = Vs/4z \tag{1}$$

HSVR is primarily used for seismic hazard mitigation through seismic response analysis of the waste deposition site that could allow prediction of seismic displacements of cover sliding (Zekkos 2005). Applications of the HSVR method, including studies for waste management sites, have been implemented by multiple scientists (Malte Ibs-von and Wohlenberg 1999; Delgado et al. 2000a, b; Parolai et al. 2001; Parolai 2002; Soupios et al. 2007a; Karagoz et al. 2015).

Time Domain (TEM or TDEM) and Frequency Domain (FDEM) Electromagnetic Method

The TDEM method measures the electrical conductivity of soils and rocks by inducing pulsating currents in the ground with a transmitter coil and monitoring the decay of the induced current over time with a separate receiver coil (ASTM 2001); in the frequency domain variant (FDEM) the magnitude and phase of an induced electromagnetic current is measured (ASTM 2001; Reynolds 2011). TDEM and FDEM measurements are ideal to map lateral changes in subsurface conductivity, determine depth and thickness of natural geologic and hydrologic layers, detect and map landfill leachate plumes, monitor seepage from brine pits and saltwater intrusion and determine fracture orientation (Chongo et al. 2015; Soupios et al. 2015; Kourgialas et al. 2016). The advantages of TDEM and FDEM methods are the good lateral and vertical resolution and the extended depth range (from a few meters to 1 km). FDEM methods can provide more accurate subsurface conductivity images when constrained by ERI surveys (Minsley et al. 2012; Briggs et al. 2016). Both methods though have the following limitations, (a) deep measurements require a large transmitter coil for which space may not be readily available, (b) susceptibility to interference from nearby metal pipes, cables, fences, vehicles, and induced noise from power lines, and (c) the effectiveness of electromagnetic measurements decreases at very low conductivities.

Ground Penetration Radar (GPR)

Ground penetrating radar (GPR) uses high-frequency electromagnetic waves to acquire subsurface information (ASTM 2005). Energy is radiated downward into the ground from a transmitter and the reflected/refracted energy is sensed by a receiving antenna. The reflected signals can produce continuous cross-sectional profiles of the shallow subsurface. Reflections of the radar wave occur where there is a change in the dielectric constant or electrical conductivity between two materials. Changes in conductivity and in dielectric properties can be the result of changes in hydrogeology, geology, moisture content, and presence of void spaces; large changes in dielectric properties often exist between geologic materials and man-made structures such as buried utilities or underground tanks. As a result, GPR can be used to provide detailed images of subsurface structures, to map buried waste, and contaminant or saline plumes (Chira Oliva et al. 2015; Iwalewa and Makkawi 2015; Wang et al. 2015; Wijewardana et al. 2015). Measurements are relatively easy to make, allowing for fast spatial coverage, and provide relatively high resolution (depending on the antennae used and ground properties). One important limitation of the method is that the penetration depth in conductive materials (>20 mS/m) such as silts and clays is very limited.

Radio/Audio-Magnetotelluric Methods (RMT/AMT)

The RMT method is an extension of the well-known very-low frequency (VLF) technique to higher frequencies (Reynolds 2011). RMT uses radio transmitters in the frequency range between 10 and 300 kHz, sometimes extended to 1 MHz. The RMT method has been used for waste disposal site characterization (Tezkan et al. 1996, 2000; Zacher et al. 1996; Newman et al. 2003). RMT surveys have been successfully used for waste site characterization, e.g., (Tezkan 1999; Tezkan et al. 2000; Newman et al. 2003). RMT data can be inverted to provide 2D and 3D subsurface reconstruction, with variety of approaches (e.g., the L2 and Laplacian norm of model parameters); generally, a priory information used during the inversion process can help produce more accurate results, especially with depth (Newman et al. 2003).

Magnetic Susceptibility (MS)

Magnetic susceptibility (MS) describes the magnetization of materials under an externally applied field, per unit of the applied field (Huliselan et al. 2010; Bijaksana et al. 2013; Kim et al. 2015). Strictly speaking, MS is the ratio of the material magnetization to the strength of the applied magnetic field; MS typically refers to the volume affected by the external field, and it depends on the magnetic properties of the components. Based on their MS properties materials can be categorized in paramagnetic, ferromagnetic, or diamagnetic. In most sediments elevated magnetic susceptibility values indicate the presence of iron-rich materials. The use of both total magnetic field and magnetic susceptibility measurements allow the detection of ferromagnetic minerals such as pyrite (FeS_2). The measurement of the three orthogonal magnetic field components (magnetic zones), represent the local value of the normal ambient field of the Earth as modified by the remnant magnetization of adjacent sediments (Prezzi et al. 2005; Almadani et al. 2015). The identification of such magnetic zones indicates layers that may have higher permeability and therefore may be potential flow paths for groundwater. Recently MS has been suggested as a proxy for characterizing and monitoring hydrocarbon degradation processes (Atekwana et al. 2014; Jobin et al. 2016).

Borehole Applications

Geophysical methods can be applied on the surface, but also in boreholes (Rubin and Hubbard 2005; Reynolds 2011). Borehole geophysical methods provide continuous profiles, point measurements at discrete depths in a borehole, and cross-borehole tomographic images. Borehole methods provide detailed information with depth without resolution loss as with surface application. Borehole geophysical methods can be performed in single boreholes, in cross hole, and even in multi-borehole configuration.

Crosshole/Downhole Seismic (CS/DS)

For successful remediation of contaminated site an accurate characterization of subsurface geology is required. Currently, the established method for acquiring such information is well log data. Well logs provide very accurate and detailed information on a site's subsurface but are spatially limited, and highly invasive (in contaminated sites well pose the risk of extending contamination problems). Cross hole seismic imaging provides a means to extend geological characterization beyond the borehole, typically in the plane(s) between boreholes (Binley et al. 2002b; Delgado et al. 2002; Rubin and Hubbard 2005; Baker et al. 2015; Sahadewa et al. 2015; Anbazhagan et al. 2016; Dantas and Medeiros 2016). Crosshole and downhole seismic imaging involves the deployment of seismic source and receivers in two—or more—boreholes and/or on the surface, surrounding the area to be imaged. Commonly the seismic wave travel times are measured and then are processed (inverted) using tomographic approaches. Borehole seismic method can be used for characterization, or in time lapse mode for monitoring the progress of remediation projects.

Electrical Resistivity Tomography (ERT)

Surface electrical surveys can be extended by placing electrodes in a single, or multiple, boreholes (Rubin and Hubbard 2005). The electrodes can either be placed permanently, or downloaded for each use, provided the borehole is uncased or PVC lined with sufficient slotted/open intervals. When two or more boreholes are used the suggested term is electrical resistivity tomography (ERT).

ERT offers certain advantages over ERI such as high resolution with depth, and not requirement for surface access (e.g., below buildings). The disadvantages are that boreholes are required, survey area is constrained by the boreholes, and data acquisition and processing might be more complicated and challenging. Very often surface electrodes are combined with borehole electrodes to provide more accurate image of the subsurface.

Cross-borehole ERT has been successfully used for a wide array of applications in different environments. One of the earliest examples of hydrological applications of ERT is Daily et al. (1992) in a study of vadose zone moisture migration due to application of a tracer. Other examples of unsaturated zone studies using ERT demonstrated how three- and two-dimensional ERT can be used successfully to monitor changes in moisture content in unsaturated sandstone (Kemna et al. 2000, 2004; Binley et al. 2002a; French et al. 2002).

Borehole Electromagnetic (BEM)

The electrical properties of the subsurface can also be investigated through EM induction in borehole configurations (BEM). As with ERT applications, BEM can be applied to single borehole configuration as well as in tomographic mode between two and more boreholes. Single borehole logging, the most common approach, can provide information on the vertical distribution of conductivity with high resolution (Williams et al. 1993).

Typical application of BEM include detection of [a] screened intervals in groundwater monitoring wells, [b] conductivity changes outside of cased wells, and [c] plume monitoring in the vadose zone (Dawson 2002). Although the BEM method does not offer the resolution of ERI/ERT, it does not rely on direct contact with the formation, is not limited to fluid filled boreholes and can provide rapid results.

Application in Landfills

Electrical Methods

Geoelectrical imaging techniques can be utilized for a variety of characterization and monitoring purposes in landfills and waste management processes (Klefstad et al. 1977; Aristodemou and Thomas-Betts 2000; Cassiani et al. 2006; Chambers et al. 2006; Grellier et al. 2008; Carlo et al. 2013; Vargemezis et al. 2015; Yin et al. 2015; Abdulrahman et al. 2016; Ntarlagiannis et al. 2016). Regulated waste management, and unregulated waste dumping, will result in significant changes to subsurface electrical properties that can be measured with geoelectrical techniques.

As it pertains to landfills, geoelectrical methods can provide subsurface images of conductivity distribution that can be linked (interpreted) to parameters of interest such as contaminant concentration gradients (e.g., leaks), structural integrity (e.g., fractures, clay caps), geological description of the surrounding area (e.g., faults, lithology), and extent of buried waste (Dawson 2002; Soupios et al. 2007a; Kemna et al. 2012; Revil et al. 2012b; Genelle et al. 2014; Tsourlos et al. 2014; Vargemezis et al. 2015). Furthermore, geoelectrical methods can be used as a cost-efficient method for long-term monitoring (Heenan et al. 2015), either for leachate leaks or for biogas production (Soupios et al. 2007a).

Leachate Monitoring

Geoelectrical methods have been routinely used to monitor for leachate leaks in landfills. Leachate monitoring involves [a] liner leak detection, and [b] contaminant plume monitoring methods. Liquid waste in landfills is generally associated with high ion concentrations, resulting in high conductivities. The conductive leachate can then be used to identify holes in the bottom liners and any movement of the plume past the landfill boundaries (Slater et al. 1997; Cassiani et al. 2006; Clément et al. 2010; Tsourlos et al. 2014).

Bottom liners in landfills are used to keep the waste from interacting with the surrounding environment, effectively electrically isolating the landfill. The basic principle of leak detection in landfills relies on this property: electrical isolation of the landfill as long as there are no holes in the liner. The integrity of the liner can be simply investigated by testing if current can move past the liner (Binley et al. 1997; Carlo et al. 2013; Tsourlos et al. 2014). There are two common methods used to test for leaks in landfills: [a] the roving (moving) electrode method (Fig. 1), and [b] permanent monitoring installation (Reynolds 2011). In the former case a pair of electrodes is used to inject current placed outside the liner, and a pair is used to detect any signal generated inside the landfill (Tsourlos et al. 2014) (Fig. 1). Response is expected only if there is a hole in the liner, allowing the current to penetrate; by moving the measuring pair of electrodes the location of the hole can be located (Fig. 1). Variations of the method include the use of one roving measurement electrode, with the second being placed outside the landfill (Reynolds 2011). In many newer landfills a monitoring network is established by permanently installing electrodes below the bottom liner; in this case the potential is being measured between different electrodes (no need for roving electrodes) and the monitoring can be continuous, and is usually automated (Reynolds 2011). ERI leak monitoring is not only limited to bottom liners, but can be successfully used to map landfill caps, and identify any damage (Genelle et al. 2014).

Landfill leachate (municipal, and mine waste piles) is generally very conductive due to elevated total dissolved solids (TDS) and high ion concentration. As a result, leachate plumes are prime targets for geoelectrical methods. Indeed, ERI has been routinely used for monitoring of leaks in active, closed, and abandoned landfills

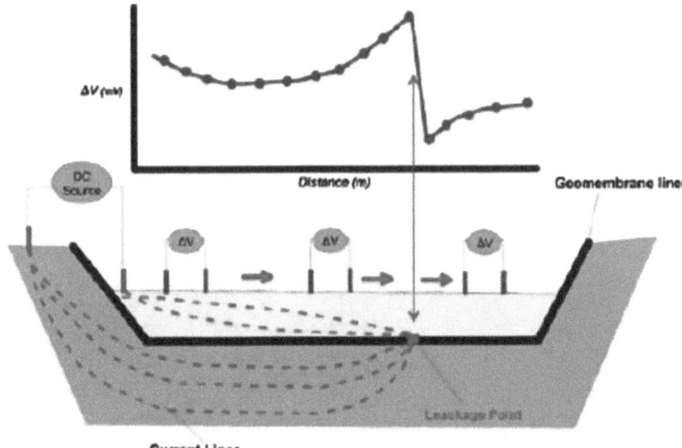

Fig. 1 Schematic showing the basic principle for liner hole detection in landfills (from Tsourlos et al. 2014)

(Greenhouse and Harris 1983; Greenhouse and Monier-Williams 1985; Bevc and Morrison 1991; Blum 1998; Chambers et al. 2006; Soupios et al. 2007a, b; Tsourlos et al. 2014; Ntarlagiannis et al. 2016). Tsourlos et al. (2014) used ERI over a closed landfill and was able to identify leaks through fractured rock; they were successful in reconstructing the leaking fracture zones based on intensive ERI characterization (Fig. 2). Ntarlagiannis et al. (2016) recently showed that ERI and IP can be successfully utilized for temporal leachate monitoring; they monitored an olive oil mill waste deposition pit for 15 months showing that conductive leachate is leaking from the pit at times of high waste load. Furthermore, ERI can be used to monitor water fluxes in landfills including leachate recirculation in bioreactor landfills (Guérin et al. 2004; Grellier et al. 2008); in the latter case ERI appears to be a very valuable tool for optimizing leachate recirculation, hence improving bioreactor landfill performance (Rosqvist and Destouni 2000; Barlaz and Reinhart 2004; Guérin et al. 2004; Rosqvist et al. 2005; Grellier et al. 2008; Valois et al. 2016).

ERI and IP are also used to reconstruct the subsurface structure of landfills. ERI has been successfully used to map the spatial distribution of parameters of interest, such as TDS and chloride content (Meju 2000b; Gazoty et al. 2012). Furthermore, the inner structure of the landfill waste can be described with electrical methods (Soupios et al. 2007a, b; Tsourlos et al. 2014; Çınar et al. 2016) (Fig. 3). Additionally, information on the structural integrity and the surrounding geology can be acquired. Such information can be used during the landfill design, and construction, but also for monitoring purposes during operation and post closure (Gazoty et al. 2012). It should be noted that although commonly only the ERI method is used since it is easier to apply (Ntarlagiannis et al. 2016), IP can provide a wealth of additional information, sometimes critical (Weller et al. 2000; Placencia-Gómez et al. 2010, 2014; Villain et al. 2011; Gazoty et al. 2012; Günther and Martin 2016; Ntarlagiannis et al. 2016).

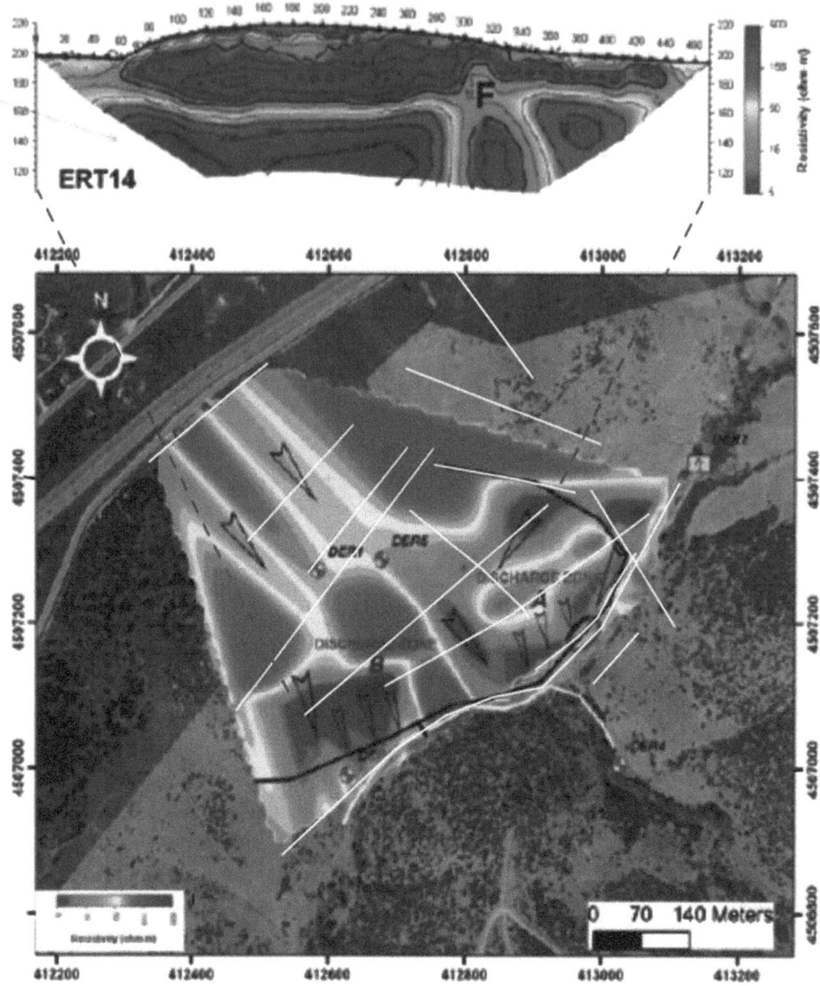

Fig. 2 2D and 3D resistivity images from and ERI survey over a landfill. ERI was successful to identify fractures (*black arrows*) that allow the landfill leachate to pollute nearby streams. *White lines* show the location of the ERI surface survey (*red dotted lines* show the location of the 2D *panel*). Modified from Tsourlos et al. 2014

The SP method's sensitivity to groundwater movement—including contaminant plumes, direct link to redox gradients typically found at plume boundaries, and easy and cost-efficient application make it an attractive option for landfill monitoring and characterization (Fig. 4). Many researchers identified SP as a potential contaminant plume monitoring tool (Weigel 1989; Hämmann et al. 1997; Buselli and Lu 2001; Nimmer 2002; Nyquist and Corry 2002; Naudet 2003a; Revil et al. 2003; Maineult et al. 2006; Arora et al. 2007). If properly applied SP data can provide information on groundwater movement, identifying the flow direction of possible contaminant

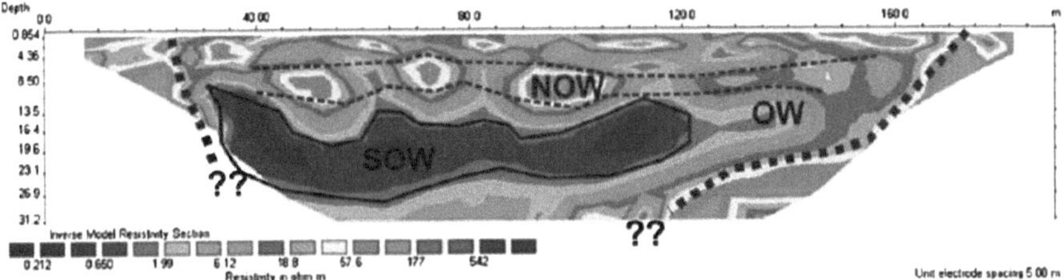

Fig. 3 Inverted ERI image showing resistivity changes that allow subsurface waste characterization. SOW (saturated organic waste), OW (organic waste), and NOW (non-organic waste) highlight areas interpreted to have saturated organic, unsaturated organic, and inorganic waste respectively. From Soupios et al. (2007a)

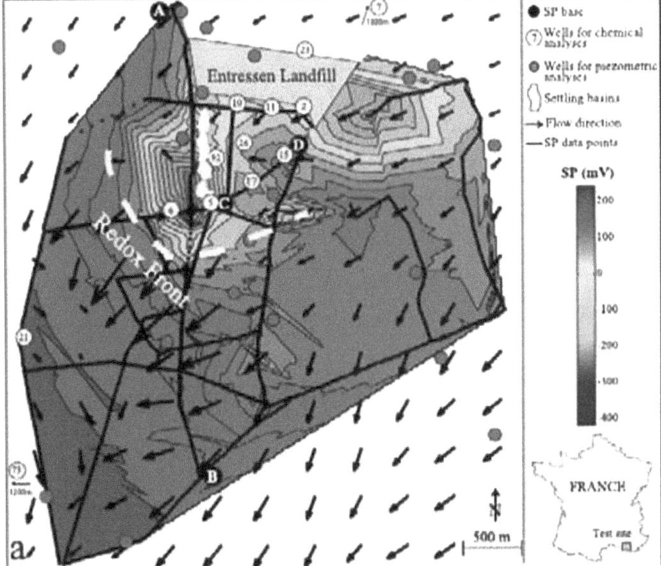

Fig. 4 SP map around the Entressen landfill; SP represents residual values after the electrokinetic contribution has been removed. SP, in agreement with redox measurements, shows a sharp anomaly at the contaminated plume front (from Naudet 2003a)

plumes. Moreover, residual SP signals have been linked to redox processes in contaminated areas, including landfills (Nyquist and Corry 2002; Naudet 2003a); this observed link is an area of active research where quantitative interpretation is the objective. One very promising model is the biogeobattery model, introduced over a decade ago that provides a direct link between microbial degradation processes, and observed SP signals (Naudet 2003a; Arora et al. 2007; Revil et al. 2010).

Summarizing, geoelectrical methods can be used to convincingly demonstrate how a proposed landfill will be sited, designed, constructed, operated, closed, and post-closure cared in order to protect the groundwater resources, public health, and the environment (Soupios et al. 2007a, b).

Electromagnetic (EM) Methods

As discussed earlier degradation of domestic putrescible solid waste, and the accumulation of liquid wastes into landfills can generate conductive leachate that fills the pore spaces; this conductive leachate can be imaged with EM methods such as FDEM and TDEM (Hutchinson 1995). In general, EM surveys are used for the rapid characterization of landfill's boundaries (Hutchinson and Barta 2000; Pellerin 2002; Monteiro Santos et al. 2006; Belmonte-jiménez et al. 2014; Wang et al. 2015; Ammar and Kruse 2016; Jodeiri Shokri et al. 2016), mapping different waste (organic, inorganic, etc.) (Mack and Maus 1986; Stenson 1988; McQuown et al. 1991; Bisdorf and Lucius 1999; Stanton and Schrader 2001; Soupios et al. 2005, 2007a, b) and detection of leachate contaminant plumes (Mack and Maus 1986; Walther et al. 1986; Hall and Pasicznyk 1987; Fawcett 1989; Russell 1990; Olhoeft and King 1991). Joint processing of geophysical data can lead to improved subsurface characterization; recently it was shown that subsurface reconstruction from FDEM can be enhanced when constrained by ERI data collected in a very small part of the surveyed area (Minsley et al. 2012; Briggs et al. 2016). Based on Hutchinson and Barta (2000), there is a linear relationship between measured terrain conductivity and waste thickness; this relationship can be used to estimate the bulk waste volume in a landfill.

Fig. 5 The conductive anomaly (leachate) below the Argos landfill as identified by TEM survey

The depth of buried waste tends to increase with landfill operation. This process leads to increasing compaction of buried waste, which in turn changes the subsurface characteristics of the landfill (e.g. pore space and saturation). GPR is sensitive to such changes, and consequently can be used for waste age classification (Splajt et al. 2003). As other EM methods, GPR can be used for contaminant plume identification and monitoring (Davis and Annan 1989; Scaife and Annan 1991; Annan 1992; Nobes 1996; Sauck et al. 1998; Green et al. 1999; Atekwana et al. 2000; Sauck 2000; Orlando and Marchesi 2001; Porsani et al. 2004). Pujari et al. (2007) jointly used EM and GPR to map the subsurface of a landfill; they were successful in mapping clay depressions, and conductive pathways in the underlying limestone.

The TEM method has been used in environmental and hydrogeologic studies over the last couple of decades. In 2013, the TEM method was used to study the hydrogeological properties of the Argolis basin in Peloponnesus (Greece). One TEM profile was performed over the main waste disposal site of the city of Argos; this profile revealed a subsurface conductive feature at 25 m depth, extending over 3 km (Fig. 5). This conductive anomaly was interpreted as contaminated leachate, that was later confirmed by direct sampling and geochemical analysis.

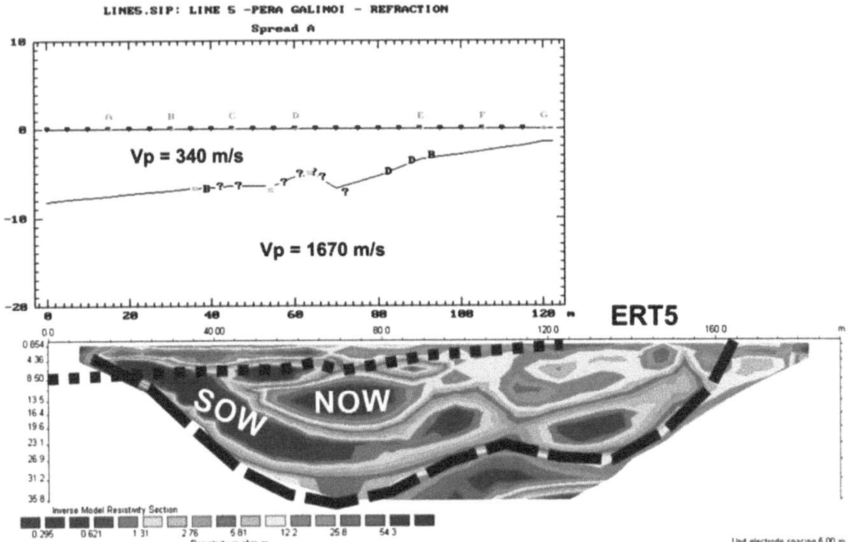

Fig. 6 Comparison of the refraction seismic (*upper*) with the ERI image (*lower*). The results are in good agreement (Soupios et al. 2007a)

Seismic Methods

The application of seismic methods can provide valuable information for the subsurface structure of a landfill. Seismic data interpretation should be performed with extra caution due to the heterogeneity that usually characterizes landfills. Common applications include characterization of the structural integrity and local geology, and mapping of the lateral continuity of buried waste (Rodriguez 1987; Slaine et al. 1990; Boyce et al. 1995; Doll et al. 1996; Cardarelli and Bernabini 1997; Doll 1998; Granda and Cambero 1998; Lanz et al. 1998; Green et al. 1999; Murray et al. 1999; De Iaco et al. 2003; Soupios et al. 2007a, b). Soupios et al. (2007a, b) applied an integrated suite of geophysical methods to characterize a landfill. The collected seismic and electrical data are in very good agreement as evidence in Fig. 6. The subsurface structure is characterized by an upper layer

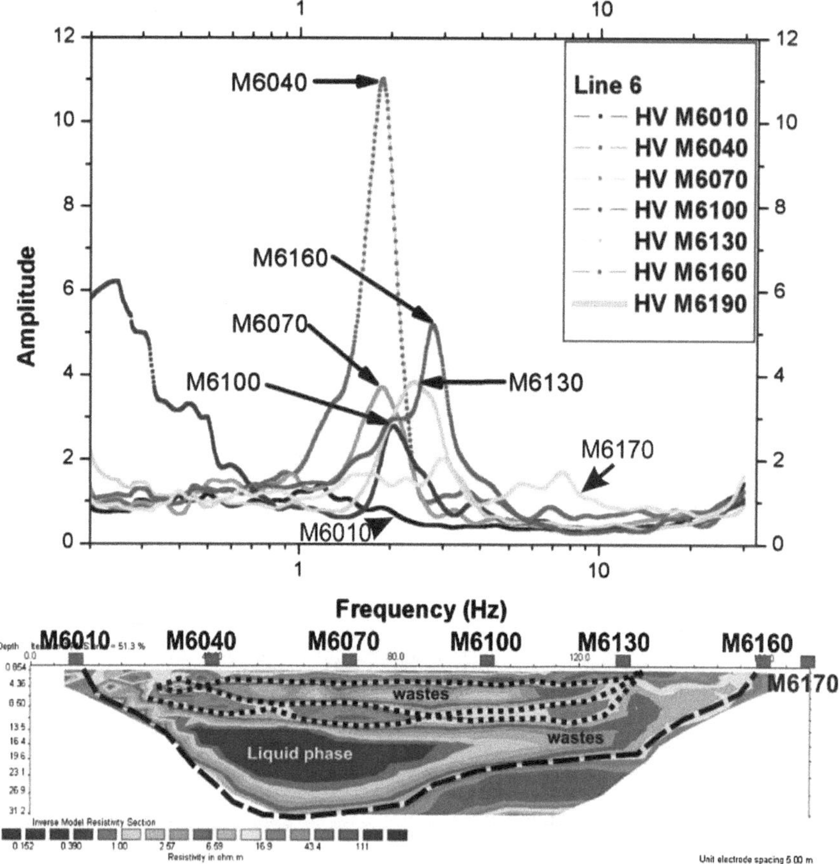

Fig. 7 HVSR measurements show a characteristic frequency peak clearly associated with waste thickness. From Soupios et al. (2007a)

($\sim$5 m thickness) with high resistivity/low velocity characteristics; below the top layer, and up to 20 m depth, we can observe a low resistivity/high velocity layer consistent with saturated conductive waste.

Possible faults and fracture zones may act as pathways for groundwater and contaminant transport to deeper horizons and the aquifer. In most cases landfills are characterized by much lower velocities (P wave velocity may range from 180 to over 700 m/s) than the surrounding sediments (Knight et al. 1978; Calkin 1989; Sharma et al. 1990), thus the seismic contrast is high and seismic refraction methods are a good choice for outlining their borders. There are only few successful applications of seismic reflection surveys across landfills (Pasasa et al. 1998). The main reasons that seismic reflection techniques may fail to be applied in landfill characterization are: (a) high levels of scattering and anelastic attenuation cause unconsolidated wastes to be generally poor transmitters of seismic waves, (b) source-generated noise (i.e., direct, refracted, guided, and surface waves) may mask shallow reflections (Robertsson et al. 1996; Roth et al. 1998), and (c) strong lateral velocity variations (inhomogeneous environment) may inhibit the recording of hyperbolic-shaped events, making identification of reflections difficult. Recently, Konstantaki et al. (2015), Konstantaki (2016) managed to apply successfully S-wave reflection seismic to map in high-resolution heterogeneities in a landfill and to estimate the density changes from S-wave velocity analysis using specific acquisition parameters and special processing steps during the velocity analysis (Konstantaki 2016).

Joint acquisition of multiple geophysical data, along with direct sampling and geotechnical data can enhance landfill characterization. Soupios et al. (2005, 2007a, 2008) highlights the synergistic use of multiple geophysical methods by interpreting shear velocity data, using Eq. 1, estimating the needed parameters with ERI (basement depth (z)) and HSVR (main resonance frequency). Shear velocities when then used to develop the stiffness model of the landfill, needed for calculating the response of the landfill in earthquakes. Figure 7 clearly shows that the estimated amplitude is proportional to the waste composition and thickness. The first (M6010) and the last (M6170) measurements have almost flat response (located at the outcrop of the bedrock), while all the other measurements (M6040-M6160) have resonance frequency that appears as a single pick in the spectrum. The amplitude (amplification) of M6040-M6160 sites depends on the thickness and composition of the underlying waste.

Information such as Vs model and/or amplification of a site can be safely used by the engineers for a detailed estimation of seismic site response analysis and the prediction and protection of seismic displacements of a landfill (Kavazanjian and Matasovic 1995; Kavazanjian et al. 1996; Augello et al. 1998; Matasovic and Kavazanjian 1998; Zekkos 2005) or other correlated environmental problems (landfill failure and uncontrolled release of the contaminants). A successful example is described in Soupios et al. (2007b) where the synergistic use of ERI and HVSR identified a subsurface karstic void in a location where the leachate collecting tanks (large static load) where to be placed; the engineers updated the landfill geotechnical design based on the geophysical information. The application of ambient noise (HVSR) measurements in similar cases is very promising since this method is cost-effective, nondestructive, and easy to apply (Soupios et al. 2007a, b).

Concluding Statement

Currently used methods for landfill characterization and monitoring typically rely on a network of geotechnical and/or monitoring wells; this approach is expensive, invasive, and spatially and temporally limited. The complementary use of noninvasive geophysical methods offers unique advantages that could significantly improve waste management practices. We should highlight that geophysical methods are not intended to replace well monitoring/characterization, but to enhance and optimize existing protocols (including cost reduction).

All geophysical methods discussed can be used during the design, construction, operation, and post-closure monitoring of a landfill site. Synergistic use of the methods can provide information on almost every aspect of a waste management site, from the geological setting and the structural integrity, to leachate monitoring and recirculation, to leak detection, and liner and cap integrity monitoring. Care should be taken on geophysical data interpretation, and joint processing of geotechnical, geochemical, and geophysical is recommended.

References

Abdulrahman A, Nawawi M, Saad R, Abu-Rizaiza AS, Yusoff MS, Khalil AE, Ishola KS (2016) Characterization of active and closed landfill sites using 2D resistivity/IP imaging: case studies in Penang, Malaysia. Environ Earth Sci 75(4):347. doi:10.1007/s12665-015-5003-5

Almadani S, Ibrahim E, Abdelrahman K, Al-Bassam A, Al-Shmrani A (2015) Magnetic and seismic refraction survey for site investigation of an urban expansion site in Abha District, Southwest Saudi Arabia. Arab J Geosci 8(4):2299–2312. doi:10.1007/s12517-014-1342-x

Ammar AI, Kruse SE (2016) Resistivity soundings and VLF profiles for siting groundwater wells in a fractured basement aquifer in the Arabian Shield, Saudi Arabia. J African Earth Sci 116:56–67. doi:10.1016/j.jafrearsci.2015.12.020

Anbazhagan P, SivakumarBabu G, Lakshmikanthan P, VivekAnand K (2016) Seismic characterization and dynamic site response of a municipal solid waste landfill in Bangalore, India. Waste Manag Res 34(3):205–213. doi:10.1177/0734242X15622814

Annan AP (1992) Ground penetrating radar. Mississauga, ON, Canada

Aristodemou E, Thomas-Betts A (2000) DC resistivity and induced polarisation investigations at a waste disposal site and its environments. J Appl Geophys 44(2–3):275–302. doi:10.1016/S0926-9851(99)00022-1

Arora T, Linde N, Revil A, Castermant J (2007) Non-intrusive characterization of the redox potential of landfill leachate plumes from self-potential data. J Contam Hydrol 92(3–4):274–292. doi:10.1016/j.jconhyd.2007.01.018

ASTM (2001) Standard guide for using the frequency domain electromagnetic method for subsurface investigations, ASTM D6639-01

ASTM (2005) Standard guide for using the surface ground penetrating radar method for subsurface investigation ASTM D6432-99

ASTM (2006) Standard guide for using the seismic refraction method for subsurface investigation. J Appl Geophys 44(2–3):167–180. doi:10.1016/S0926-9851(98)00033-0

Atekwana EA, Sauck W, Werkema WA Jr (2000) Investigations of geoelectrical signatures at a hydrocarbon contaminated site. J Appl Geophys 44(2–3):167–180. doi:10.1016/S0926-9851(98)00033-0

Atekwana EA, Mewafy FM, Abdel Aal G, Werkema DD, Revil A, Slater LD (2014) High-resolution magnetic susceptibility measurements for investigating magnetic mineral formation during microbial mediated iron reduction. J Geophys Res Biogeosci 119(1):80–94. doi:10.1002/2013JG002414

Augello A, Bray J, Abrahamson N, Seed R (1998) Dynamic properties of solid waste based on back-analysis of OII landfill. ASCE J Geotech Geoenviron Eng 124(3):211–222

Baker H, Gabr A, Djeddi M (2015) Geophysical and geotechnical techniques: complementary tools in studying subsurface features. In: International conference on engineering geophysics, Al Ain, United Arab Emirates, 15–18 Nov 2015, Society of Exploration Geophysicists, pp 2–5

Barlaz MA, Reinhart D (2004) Bioreactor landfills: progress continues. Waste Manag 24(9):859–860. doi:10.1016/j.wasman.2004.09.001

Belghazal H, Piga C, Loddo F, El Messari JS, Touhami AO (2013) Geophysical surveys for the characterization of landfills. Int J Innov Appl Stud 4(2):254–263

Belmonte-jiménez SI, Bortolotti-villalobos A, Campos-enríquez JÓ, Pérez-flores MA, Delgado-rodríguez O, Ensenada-tijuana C, Playitas Z, California B, México CP (2014) Electromagnetic methods application for characterizing a site. Rev Int Contam Ambie 30(3):317–329

Benson RC, Yuhr LB (2016) Engineering measurements and monitoring. Site characterization in karst and pseudokarst terraines. Springer, Netherlands, Dordrecht, pp 265–273

Benson RR, Glaccum A, Noel M (1988) Geophysical techniques for sensing buried wastes and waste migration. National Water Well Association, Dublin

Bevc D, Morrison H (1991) Borehole-to-surface electrical resistivity monitoring of a salt water injection experiment. Geophysics 56(6):769–777

Bijaksana S, Huliselan E, Safiuddin LO, Fitriani D, Tamuntuan G, Agustine E (2013) Rock magnetic methods in soil and environmental studies: fundamentals and case studies. Procedia Earth Planet Sci 6:8–13. doi:10.1016/j.proeps.2013.01.001

Binley A, Daily W, Ramirez A (1997) Detecting leaks from environmental barriers using electrical current imaging. J Environ Eng Geophys 2(1):11–19. doi:10.4133/JEEG2.1.11

Binley A, Winship P, West LJJ, Pokar M, Middleton R (2002a) Seasonal variation of moisture content in unsaturated sandstone inferred from borehole radar and resistivity profiles. J Hydrol 267(3–4):160–172. doi:10.1016/S0022-1694(02)00147-6

Binley A, Cassiani G, Middleton R, Winship P (2002b) Vadose zone flow model parameterisation using cross-borehole radar and resistivity imaging. J Hydrol 267(3–4):147–159. doi:10.1016/S0022-1694(02)00146-4

Bisdorf RJ, Lucius JE (1999) Mapping the Norman, Oklahoma landfill contamination plume using electrical geophysics. In: Morganwalp HT, Buxton DW (eds) US geological survey toxic substances hydrology program proceedings of the technical meeting, US Geological Survey Water-Resources Investigations Report 99-4018-C, Charleston, South Carolina, pp 579–584

Blum R (1998) Geoelectrical mapping and groundwater contamination. In: Merkler GP (ed) Detection of subsurface flow phenomena. Springer, Berlin, pp 253–260

Boyce JI, Eyles N, Pugin A (1995) Seismic reflection, borehole and outcrop geometry of late wisconsin tills at a proposed landfill near Toronto, Ontario. Can J Earth Sci 32(9):1331–1349. doi:10.1139/e95-108

Briggs MA, Campbell S, Nolan J, Walvoord MA, Ntarlagiannis D, Day-Lewis FD, Lane JW (2016) Surface geophysical methods for characterizing the active layer and shallow permafrost features. Permafr Periglac Process doi:10.1002/ppp.1893

Buselli G, Lu K (2001) Groundwater contamination monitoring with multichannel electrical and electromagnetic methods. J Appl Geophys 48(1):11–23. doi:10.1016/S0926-9851(01)00055-6

Calkin S (1989) A shallow seismic refraction survey of the Mallard North Landfill—Hanover Park. Northern Illinois Univ, IL, US

Cardarelli E, Bernabini M (1997) Two case studies of the determination of parameters of urban waste dumps. J Appl Geophys 36(4):167–174. doi:10.1016/S0926-9851(96)00056-0

De Carlo L et al (2013) Characterization of a dismissed land fi ll via electrical resistivity tomography and mise-à-la-masse method. J Appl Geophys 98:1–10. doi:10.1016/j.jappgeo.2013.07.010

Carpenter PJ, Calkin SF, Kaufmann RS (1991) Assessing a fractured landfill cover using electrical resistivity and seismic refraction techniques. Geophysics 56(11):1896–1904. doi:10.1190/1.1443001

Cassiani G, Bruno V, Villa A, Fusi N, Binley A (2006) A saline trace test monitored via time-lapse surface electrical resistivity tomography. J Appl Geophys 59(3):244–259. doi:10.1016/j.jappgeo.2005.10.007

Chambers JE, Kuras O, Meldrum PI, Ogilvy RD, Hollands J (2006) Electrical resistivity tomography applied to geologic, hydrogeologic, and engineering investigations at a former waste-disposal site. Geophysics 71(6):B231–B239. doi:10.1190/1.2360184

Chira Oliva P, Barbalho Pires D, Ribeiro Cruz J (2015) Environmental study of the Bragança City landfill (Brazil) applying ground penetrating radar. EAGE, 21st European Meeting of Environmental and Engineering Geophysics. doi: 10.3997/2214-4609.201413820

Chongo M, Christiansen AV, Fiandaca G, Nyambe IA, Larsen F, Bauer-Gottwein P (2015) Mapping localised freshwater anomalies in the brackish paleo-lake sediments of the Machile-Zambezi Basin with transient electromagnetic sounding, geoelectrical imaging and induced polarisation. J Appl Geophys 123:81–92. doi:10.1016/j.jappgeo.2015.10.002

Çınar H, Altundaş S, Ersoy E, Bak K, Bayrak N (2016) Application of two geophysical methods to characterize a former waste disposal site of the Trabzon-Moloz district in Turkey. Environ Earth Sci 75(1):52. doi:10.1007/s12665-015-4839-z

Clément R, Descloitres M, Günther T, Oxarango L, Morra C, Laurent J-P, Gourc J-P (2010) Improvement of electrical resistivity tomography for leachate injection monitoring. Waste Manag 30(3):452–464. doi:10.1016/j.wasman.2009.10.002

Daily W, Ramirez A, LaBrecque D, Nitao J (1992) Electrical resistivity tomography of vadose water movement. Water Resour Res 28(5):1429–1442. doi:10.1029/91WR03087

Dantas RRS, Medeiros WE (2016) Resolution in crosswell travel time tomography: the dependence on illumination. Geophysics 81(1):W1–W12. doi:10.1190/geo2015-0119.1

Davis JL, Annan AP (1989) Ground-penetrating radar for high-resolution mapping of soil and rock stratigraphy 1. Geophys Prospect 37(5):531–551. doi:10.1111/j.1365-2478.1989.tb02221.x

Dawson C, Lane J Jr (2002) Integrated geophysical characterization of the Winthrop landfill southern flow path, Winthrop, Maine. … Appl Geophys … 1–22

Delgado J, López Casado C, Estévez A, Giner J, Cuenca A, Molina S (2000a) Mapping soft soils in the Segura river valley (SE Spain): a case study of microtremors as an exploration tool. J Appl Geophys 45(1):19–32. doi:10.1016/S0926-9851(00)00016-1

Delgado J, López Casado C, Giner J, Estévez A, Cuenca A, Molina S (2000b) Microtremors as a geophysical exploration tool: applications and limitations. Pure Appl Geophys 157(9):1445–1462. doi:10.1007/PL00001128

Delgado J, Alfaro P, Galindo-Zaldivar J, Jabaloy A, López Garrido CA, Sanz de Galdeano C (2002) Structure of the Padul-Nigüelas Basin (S Spain) from H/v ratios of ambient noise: application of the method to study peat and coarse sediments. Pure Appl Geophys 159(11–12)–2749. doi:10.1007/s00024-002-8756-1

Doll WE (1998) Reprocessing of shallow seismic reflection data to image faults near a hazardous waste site on the Oak Ridge Reservation, Tennessee. In: Symposium application geophysics environmental engineering problems (SAGEEP), proceedings. Environmental and Engineering Geophysical Society, Denver, CO, pp 705–714

Doll WE, Miller RD, Xia J (1996) Enhancement of swept source near-surface seismic reflection data at a hazardous waste site. In: 66th Annual international meeting society of exploration geophysicists expanded abstracts, Society of Exploration Geophysics, Tulsa, OK, pp 877–879

Fawcett JD (1989) Hydrogeologic assessment, design and remediation of a shallow groundwater contaminated zone. In: Proceedings of the 3rd national outdoor action conference on aquifer restoration, ground water monitoring and geophysical methods, National Water Well Association, Orlando, pp 591–605

French HK, Hardbattle C, Binley A, Winship P, Jakobsen L (2002) Monitoring snowmelt induced unsaturated flow and transport using electrical resistivity tomography. J Hydrol 267(3–4):273–284. doi:10.1016/S0022-1694(02)00156-7

Gazoty A, Fiandaca G, Pedersen J, Auken E, Christiansen AV, Pedersen JK (2012) Application of time domain induced polarization to the mapping of lithotypes in a landfill site. Hydrol Earth Syst Sci 16(6):1793–1804. doi:10.5194/hess-16-1793-2012

Genelle F, Sirieix C, Riss J, Naudet V, Dabas M, Bégassat P (2014) Detection of landfill cover damage using geophysical methods. Near Surf Geophys 12(2036):599–611. doi:10.3997/1873-0604.2014018

Gouveia F, Lopes I, Gomes RC (2016) Deeper VS profile from joint analysis of Rayleigh wave data. Eng Geol 202:85–98. doi:10.1016/j.enggeo.2016.01.006

Granda A, Cambero JC (1998) The use of geophysical techniques for the detection and characterization of landfill in areas of urban development. In: 4th annual meeting environmental engineering geophysics society, european section, proceedings, Environmental and Engineering Geophysical Society, Lausanne, Switzerland, pp 111–114

Green A, Lanz E, Maurer H (1999) A template for geophysical investigations of small landfills. Lead Edge 18(2):248–254

Greenhouse J, Harris R (1983) Migration of contaminants in groundwater at a landfill: a case study. J Hydrol 63(1–2):177–197. doi:10.1016/0022-1694(83)90227-5

Greenhouse JP, Monier-Williams M (1985) Geophysical Monitoring of ground water contamination around waste disposal sites. Ground Water Monit Remediat 5(4):63–69. doi:10.1111/j.1745-6592.1985.tb00940.x

Greenwood W, Zekkos D, Sahadewa A (2015) Spatial variation of shear wave velocity of waste materials from surface wave measurements. J Environ Eng Geophys 20(4):287–301. doi:10.2113/JEEG20.4.287

Grellier S, Guerin R, Robain H, Bobachev A, Vermeersch F, Tabbagh A (2008) Monitoring of leachate recirculation in a bioreactor landfill by 2-D electrical resistivity imaging. J Environ Eng Geophys 13(4):351–359. doi:10.2113/JEEG13.4.351

Guérin R, Munoz ML, Aran C, Laperrelle C, Hidra M, Drouart E, Grellier S (2004) Leachate recirculation: moisture content assessment by means of a geophysical technique. Waste Manag 24(8):785–794. doi:10.1016/j.wasman.2004.03.010

Günther T, Martin T (2016) Spectral two-dimensional inversion of frequency-domain induced polarization data from a mining slag heap. J Appl Geophys 1–13. doi:10.1016/j.jappgeo.2016.01.008

Hall DW, Pasicznyk DL (1987) Application of seismic refraction and terrain conductivity methods at a ground water pollution site in North-Central New Jersey. In: Graves B, Lehr JH, Butcher K, Alcorn P, Ammerman L, Williams P, Renz M, Shelton V (eds) 1st national outdoor action conference on aquifer restoration, groundwater monitoring and geophysical methods. National Water Well Association, Las Vegas, Nevada, pp 505–524

Hämmann M, Maurer HR, Green AG, Horstmeyer H (1997) Self-Potential Image Reconstruction: Capabilities and Limitations. J Environ Eng Geophys 2(1):21–35. doi:10.4133/JEEG2.1.21

Heenan J, Slater LD, Ntarlagiannis D, Atekwana EA, Fathepure BZ, Dalvi S, Ross C, Werkema DD, Atekwana EA (2015) Electrical resistivity imaging for long-term autonomous monitoring of hydrocarbon degradation: lessons from the deepwater horizon oil spill. Geophysics 80(1):B1–B11. doi:10.1190/geo2013-0468.1

Hix K (1998) Leak detection for landfill liners: overview of tools for vadoze zone monitoring. Technical status report EPA-542-R-98-019, US Environmental Protection Agency, Washington, DC

Huliselan EK, Bijaksana S, Srigutomo W, Kardena E (2010) Scanning electron microscopy and magnetic characterization of iron oxides in solid waste landfill leachate. J Hazard Mater 179(1–3):701–708. doi:10.1016/j.jhazmat.2010.03.058

Hutchinson PJ (1995) The geology of landfills. Environl Geosci 2(1):2–14

Hutchinson PJ, Barta LS (2000) Geophysical applications to solid waste analysis. In: Zandi I WS, Mersky RL (eds) The 16th international conference on solid waste technology and management, Philadelphia, PA, USA, pp 2–78

De Iaco R, Green A, Maurer H-R, Horstmeyer H (2003) A combined seismic reflection and refraction study of a landfill and its host sediments. J Appl Geophys 52(4):139–156. doi:10.1016/S0926-9851(02)00255-0

Iwalewa TM, Makkawi MH (2015) Site characterization and risk assessment in support of the design of groundwater remediation well near a hazardous landfill. Arab J Geosci 8(3):1705–1715. doi:10.1007/s12517-014-1300-7

Jobin P, Mercier G, Blais J-F (2016) Magnetic and density characteristics of a heavily polluted soil with municipal solid waste incinerator residues: significance for remediation strategies. Int J Miner Process 149:119–126. doi:10.1016/j.minpro.2016.02.010

Jodeiri Shokri B, Doulati Ardejani F, Moradzadeh A (2016) Mapping the flow pathways and contaminants transportation around a coal washing plant using the VLF-EM, Geo-electrical and IP techniques—a case study, NE Iran. Environ Earth Sci 75(1):62. doi:10.1007/s12665-015-4776-x

Johnson TC, Wellman D (2015) Accurate modelling and inversion of electrical resistivity data in the presence of metallic infrastructure with known location and dimension. Geophys J Int 202 (2):1096–1108. doi:10.1093/gji/ggv206

Karagoz O, Chimoto K, Citak S, Ozel O, Yamanaka H, Hatayama K (2015) Estimation of shallow S-wave velocity structure and site response characteristics by microtremor array measurements in Tekirdag region. NW Turkey, Earth, Planets Sp 67(1):176. doi:10.1186/s40623-015-0320-1

Kavazanjian EJ, Matasovic N (1995) Seismic analysis of solid waste landfills. The geoenvironment 2000 specialty conference. ASCE, New Orleans, Louisiana, pp 24–26

Kavazanjian EJ, Matasovic N, Stokoe K, Bray JD (1996) In situ shear wave velocity of solid waste from surface wave measurements. In: 2nd international congress environmental geotechnics, Balkema, Osaka, Japan, pp 97–104

Kemna A, Binley A, Ramirez A, Daily W (2000) Complex resistivity tomography for environmental applications. Chem Eng J 77(1–2):11–18

Kemna A, Binley A, Slater L (2004) Crosshole IP imaging for engineering and environmental applications. Geophysics 69(1):97–107. doi:10.1190/1.1649379

Kemna A et al (2012) An overview of the spectral induced polarization method for near-surface applications. Near Surf Geophys 453–468. doi:10.3997/1873-0604.2012027

Kim J-E, Ha D-W, Kim Y-H (2015) Separation of steel slag from landfill waste for the purpose of decontamination using a superconducting magnetic separation system. IEEE Trans Appl Supercond 25(3):1–4. doi:10.1109/TASC.2014.2365954

Klefstad G, Sendlein LVA, Palmquist RC (1977) Limitations of the electrical resistivity method in landfill investigations. Ground Water 15(5):418–427. doi:10.1111/j.1745-6584.1977.tb03185.x

Knight MJ, Leonard JG, Whiteley RJ (1978) Lucas heights solid waste landfill and downstream leachate transport—a case study in environmental geology. Bull Int Assoc Eng Geol 18(1):45–64. doi:10.1007/BF02635349

Konstantaki LA (2016) Imaging and characterization of heterogeneous landfills using geophysical methods. PhD Thesis, Delft University of Technology

Konstantaki LA, Draganov D, Ghose R, Heimovaara T (2015) Seismic interferometry as a tool for improved imaging of the heterogeneities in the body of a landfill. J Appl Geophys 122:28–39. doi:10.1016/j.jappgeo.2015.08.008

Kourgialas NN, Dokou Z, Karatzas GP, Panagopoulos G, Soupios P, Vafidis A, Manoutsoglou E, Schafmeister M (2016) Saltwater intrusion in an irrigated agricultural area: combining density-dependent modeling and geophysical methods. Environ Earth Sci 75(1):15. doi:10.1007/s12665-015-4856-y

Lanz E, Maurer H, Green AG (1998) Refraction tomography over a buried waste disposal site. Geophysics 63(4):1414–1433. doi:10.1190/1.1444443

Linde N, Doetsch J, Jougnot D, Genoni O, Dürst Y, Minsley BJ, Vogt T, Pasquale N, Luster J (2011) Self-potential investigations of a gravel bar in a restored river corridor. Hydrol Earth Syst Sci 15(3):729–742. doi:10.5194/hess-15-729-2011

Mack TJ, Maus PE (1986) Detection of contaminant plumes in ground water of Long Island, New York. USGS, Water Resources Investigations Report, 86–4045

Maineult A, Bernabé Y, Ackerer P (2006) Detection of advected, reacting redox fronts from self-potential measurements. J Contam Hydrol 86(1–2):32–52. doi:10.1016/j.jconhyd.2006.02.007

Malte Ibs-von S, Wohlenberg J (1999) Microtremor measurements used to map thickness of soft sediments. Bull Seism Soc Am 89(1):250–259

Matasovic N, Kavazanjian EJ (1998) Cyclic Characterization of OII Landfill Solid Waste. J Geotech Geoenviron Eng 124(3):197–210. doi:10.1061/1090-0241

McQuown M, Becker S, Miller P, BSR, MPT, McQuown MS (1991) Subsurface characterization of a landfill using integrated geophysical techniques. In: 5th national outdoor action conference on aquifer restoration, ground water monitoring and geophysical methods, Water Well Journal Publishing, Las Vegas, NV, pp 933–946

Meju M (2000a) Environmental geophysics: the tasks ahead. J Appl Geophys 44(2–3):63–65. doi:10.1016/S0926-9851(00)00006-9

Meju MA (2000b) Geoelectrical investigation of old/abandoned, covered landfill sites in urban areas: model development with a genetic diagnosis approach. J Appl Geophys 44(2–3):115–150. doi:10.1016/S0926-9851(00)00011-2

Minsley BJ, Smith BD, Hammack R, Sams JI, Veloski G (2012) Calibration and filtering strategies for frequency domain electromagnetic data. J Appl Geophys 80:56–66. doi:10.1016/j.jappgeo.2012.01.008

Monteiro Santos FA, Mateus A, Figueiras J, Gonzalves MA (2006) Mapping groundwater contamination around a landfill facility using the VLF-EM method—a case study. J Appl Geophys 60(2):115–125. doi:10.1016/j.jappgeo.2006.01.002

Murray C, Keiswetter D, Rostosky E (1999) Seismic refraction case studies at environmental sites. In: symp applic geophys environ engin prob (SAGEEP), proceedings. Environmental and Engineering Geophysical Society, Environmental and Engineering Geophysical Society, Denver, CO, pp 235–244

Mwakanyamale K, Slater L, Binley A, Ntarlagiannis D (2012) Lithologic imaging using complex conductivity: lessons learned from the Hanford 300 Area. Geophysics 77(6):E397–E409. doi:10.1190/geo2011-0407.1

Naudet V (2003a) Relationship between self-potential (SP) signals and redox conditions in contaminated groundwater. Geophys Res Lett 30(21):2091. doi:10.1029/2003GL018096

Naudet V (2003b) Relationship between self-potential (SP) signals and redox conditions in contaminated groundwater. Geophys Res Lett 30(21):1–4. doi:10.1029/2003GL018096

Newman GA, Recher S, Tezkan B, Neubauer FM (2003) 3D inversion of a scalar radio magnetotelluric field data set. Geophysics 68(3):791–802. doi:10.1190/1.1581032

Nimmer RE (2002) Direct current and self-potential monitoring of an evolving plume in partially saturated fractured rock. J Hydrol 267(3–4):258–272. doi:10.1016/S0022-1694(02)00155-5

Nobes DC (1996) Troubled waters: environmental applications of electrical and electromagnetic methods. Surv Geophys 17(4):393–454. doi:10.1007/BF01901640

Ntarlagiannis D, Robinson J, Soupios P, Slater L (2016) Field-scale electrical geophysics over an olive oil mill waste deposition site: evaluating the information content of resistivity versus induced polarization (IP) images for delineating the spatial extent of organic contamination. J Appl Geophys 62:1–9. doi:10.1016/j.jappgeo.2016.01.017

Nyquist JE, Corry CE (2002) Self-potential: the ugly duckling of environmental geophysics. Lead Edge 21(5):446–451. doi:10.1190/1.1481251

Olhoeft GR, King TVV (1991) Mapping subsurface organic compounds noninvasively by their reactions with clays. In: The 4th toxic substances technical meeting, Monterey, CA, pp 1–18

Orlando L, Marchesi E (2001) Georadar as a tool to identify and characterize solid waste dump deposits. J Appl Geophys 48:163–174

Parolai S (2002) New Relationships between Vs, Thickness of Sediments, and Resonance Frequency Calculated by the H/V Ratio of Seismic Noise for the Cologne Area (Germany). Bull Seismol Soc Am 92(6):2521–2527. doi:10.1785/0120010248

Parolai S, Bormann P, Milkereit C (2001) Assessment of the natural frequency of the sedimentary cover in the cologne area (Germany) using noise measurements. J Earthq Eng 5(4):541–564. doi:10.1080/13632460109350405

Pasasa L, Wenzel F, Zhao P (1998) Prestack Kirchhoff depth migration of shallow seismic data. Geophysics 63(4):1241–1247. doi:10.1190/1.1444425

Pellerin L (2002) Applications of electrical and electromagnetic methods for environmental and geotechnical investigations. Surv Geophys 23(2/3):101–132. doi:10.1023/A:1015044200567

Petiau G (2000) Second generation of lead-lead chloride electrodes for geophysical applications. Pure Appl Geophys 157(3):357–382. doi:10.1007/s000240050004

Piratoba Morales G, Fenzi N (2000) Environmental impact of the deposit of solid waste of the "Aura" Bele′ m-PA (Brazil). In: 31st international geological congress, Rio de Janeiro, Brazil, p 4218

Placencia-Gómez E, Parviainen A, Hokkanen T, Loukola-Ruskeeniemi K (2010) Integrated geophysical and geochemical study on AMD generation at the Haveri Au–Cu mine tailings, SW Finland. Environ Earth Sci 61(7):1435–1447. doi:10.1007/s12665-010-0459-9

Placencia-Gómez E, Parviainen A, Slater L, Leveinen J (2014) Spectral induced polarization (SIP) response of mine tailings. J Contam Hydrol 173C:8–24. doi:10.1016/j.jconhyd.2014.12.002

Porsani JL, Filho WM, Elis VR, Shimeles F, Dourado JC, Moura HP (2004) The use of GPR and VES in delineating a contamination plume in a landfill site: a case study in SE Brazil. J Appl Geophys 55(3–4):199–209. doi:10.1016/j.jappgeo.2003.11.001

Prezzi C, Orgeira MJ, Ostera H, Vásquez CA (2005) Ground magnetic survey of a municipal solid waste landfill: pilot study in Argentina. Environ Geol 47(7):889–897. doi:10.1007/s00254-004-1198-6

Pujari PR, Pardhi P, Muduli P, Harkare P, Nanoti MV (2007) Assessment of pollution near landfill site in Nagpur, India by resistivity imaging and GPR. Environ Monit Assess 131(1–3):489–500. doi:10.1007/s10661-006-9494-0

Ramaiah BJ, Ramana GV, Kavazanjian E, Matasovic N, Bansal BK (2016) Empirical model for shear wave velocity of municipal solid waste in situ. J Geotech Geoenviron Eng 142 (1):06015012. doi:10.1061/(ASCE)GT.1943-5606.0001389

Revil a (2003) Principles of electrography applied to self-potential electrokinetic sources and hydrogeological applications. Water Resour Res 39(5):1–15. doi:10.1029/2001WR000916

Revil A, Karaoulis M, Johnson T, Kemna A (2012a) Review: some low-frequency electrical methods for subsurface characterization and monitoring in hydrogeology. Hydrogeol J 617–658. doi:10.1007/s10040-011-0819-x

Revil A, Naudet V, Nouzaret J, Pessel M (2003) Principles of electrography applied to self-potential electrokinetic sources and hydrogeological applications, Water Resour Res 39(5). doi:10.1029/2001WR000916

Revil A, Mendonça CA, Atekwana EA, Kulessa B, Hubbard SS, Bohlen KJ (2010) Understanding biogeobatteries: where geophysics meets microbiology. J Geophys Res 115:G00G02. doi:10.1029/2009JG001065

Revil A, Karaoulis M, Johnson T, Kemna A (2012b) Review: some low-frequency electrical methods for subsurface characterization and monitoring in hydrogeology. Hydrogeol J 617–658. doi:10.1007/s10040-011-0819-x

Reynolds J (2011) An introduction to applied and environmental geophysics, 2nd edn. Wiley, ISBN: 978-0-471-48535-3, p 710

Robertsson JOA, HK, GAG (1996) Source-generated noise in shallow seismic data. Eur J Environ Engin Geophys 1, 107–124

Robinson H, Gronow J (1995) A review of landfill leachate composition in the UK. In: Institute of waste management proceedings, IWM, Northampton, pp 3–8

Robinson J, Johnson T, Slater L (2015) Challenges and opportunities for fractured rock imaging using 3D cross-borehole electrical resistivity. Geophysics 80(2):E49–E61. doi:10.1190/geo2014-0138.1

Robinson J et al (2015b) Imaging pathways in fractured rock using three-dimensional electrical resistivity tomography. Groundwater. doi:10.1111/gwat.12356

Rodriguez EB (1987) Application of gravity and seismic methods in hydrogeological mapping at a landfill site in Ontario. In: 1st national outdoor action conference on aquifer restoration, ground water monitoring and geophysical methods, Assoc. of Groundwater Sci. and Eng.—National Water Well Association, Dublin, OH, pp 487–504

Rosqvist H, Destouni G (2000) Solute transport through preferential pathways in municipal solid waste. J Contam Hydrol 46(1–2):39–60. doi:10.1016/S0169-7722(00)00127-3

Rosqvist H, Dahlin T, Lindhe C (2005) Investigation of water flow in a bioreactor landfill using geoelectrical imaging techniques. In: Tenth international waste management and landfill symposium, Proceedings Sardinia 2005, S. Margherita di Pula, Cagliari, Italy

Roth M, Holliger K, Green AG (1998) Guided waves in near-surface seismic surveys. Geophys Res Lett 98:235–248

Rozycki A, Fonticiella JMR, Cuadra A (2006) Detection and evaluation of horizontal fractures in earth dams using the self-potential method 82:145–153. doi:10.1016/j.enggeo.2005.09.013

Rubin Y, Hubbard SS (eds) (2005) Hydrogeophysics. Water Science and Technology Library, Springer, Netherlands, Dordrecht

Russell GM (1990) Application of geophysical techniques for assessing groundwater contamination near a landfill at Stuart, Florida. In: The FOCUS conference on eastern regional ground water issues, NWWA, Springfield, Mass, pp 211–225

Sahadewa A, Zekkos D, Woods RD, Stokoe KH (2015) Field testing method for evaluating the small-strain shear modulus and shear modulus nonlinearity of solid waste. Geotech Test J 38 (4):20140016. doi:10.1520/GTJ20140016

Sato M, Mooney HM (1960) The electrochemical mechanism of sulfide self-potentials. Geophysics 25(1):226–249. doi:10.1190/1.1438689

Sauck WA (2000) A model for the resistivity structure of LNAPL plumes and their environs in sandy sediments. J Appl Geophys 44(2–3):151–165. doi:10.1016/S0926-9851(99)00021-X

Sauck WA, Atekwana EA, Nash MS (1998) High conductivities associated with an LNAPL plume imaged by integrated geophysical techniques. J Environ Eng Geophys 2(3):203–212

Scaife JE, Annan AP (1991) Ground penetrating radar: a powerful, high resolution tool for mining engineering and environmental problems. Sensors and Software. Internal Report, IR-59, p 24.

Sharma HD, Dukes MT, Olsen DM (1990) Field measurements of dynamic moduli and Poisson's ratios of refuse and underlying soils at a landfill site. In: Landva GD, Knowles A (eds) Geotechnics of waste fills—theory and practice, ASTM STP 1070. American Society for Testing and Materials, Philadelphia, PA, USA, pp 57–70

Shemang EM, Mickus K, Same MP (2011) Geophysical characterization of the abandoned Gaborone Landfill, Botswana: implications for abandoned landfills in arid environments. Int J Environ Protect 1(1):1–12

Slaine DD, Pehme PE, Hunter JA, Pullan SE, Greenhouse JP (1990) Mapping overburden stratigraphy at a proposed hazardous waste facility using shallow seismic reflection methods. In: Ward SH (ed) Geotechnical and environmental geophysics. Environmental and groundwater, vol II. Society of Exploration Geophysics, Tulsa, OK, pp 273–280

Slater L (2007) Near surface electrical characterization of hydraulic conductivity: from petrophysical properties to aquifer geometries—a review. Surv Geophys 28(2–3):169–197. doi:10.1007/s10712-007-9022-y

Slater L, Binley A, Daily W, Johnson R (2000) Cross-hole electrical imaging of a controlled saline tracer injection. J Appl Geophys 44(2–3):85–102. doi:10.1016/S0926-9851(00)00002-1

Slater LD, Zaidman MD, Binley A, West LJ (1997) Electrical imaging of saline tracer migration for the investigation of unsaturated zone transport mechanisms. Hydrol Earth Syst Sci 1:291–302

Soupios P, Vallianatos F, Makris J, Papadopoulos I (2005) Determination of a landfill structure using HVSR, geoelectrical and seismic tomographies. In: International workshop in geoenvironment and geotechnics, Milos Island, Greece, pp 83–90

Soupios P, Papadopoulos N, Papadopoulos I, Kouli M, Vallianatos F, Sarris A, Manios T (2007a) Application of integrated methods in mapping waste disposal areas. Environ Geol 53(3):661–675. doi:10.1007/s00254-007-0681-2

Soupios P, Papadopoulos I, Kouli M, Georgaki I, Vallianatos F, Kokinou E (2007b) Investigation of waste disposal areas using electrical methods: a case study from Chania, Crete, Greece. Environ Geol 51(7):1249–1261. doi:10.1007/s00254-006-0418-7

Soupios P, Piscitelli S, Vallianatos F, Lapenna V (2008) Contamination delineation and characterization of waste disposal sites performing integrated and innovative geophysical methods. In: Waste management research trends, vol 11, pp 221–259

Soupios P, Kourgialas N, Dokou Z, Karatzas G, Panagopoulos G, Vafidis A, Manoutsoglou E (2015) Modeling saltwater intrusion at an agricultural coastal area using geophysical methods and the FEFLOW model. Engineering geology for society and territory, vol 3. Springer International Publishing, Cham, pp 249–252

Splajt T, Ferrier G, Frostick LE (2003) Application of ground penetrating radar in mapping and monitoring landfill sites. Environ Geol 44(8):963–967. doi:10.1007/s00254-003-0839-5

Stanton GP, TP Schrader (2001) Surface geophysical investigation of a chemical waste landfill in northwestern Arkansas. In: U.S. geological survey karst interest group proceedings. Water-Resources Investigations Report 01-4011, pp 107–115

Statom RA, Thyne GD, McCray JE (2004) Temporal changes in leachate chemistry of a municipal solid waste landfill cell in Florida, USA. Environ Geol 45(7):982–991. doi:10.1007/s00254-003-0957-0

Stenson RW (1988) Electromagnetic data acquisition techniques for landfill investigations. The symposium on the application of geophysics to engineering problems. The Society of Engineering and Mineral Exploration Geophysics, Golden CO, pp 735–746

Suski B, Revil A, Titov K, Konosavsky P, Voltz M, Dagès C, Huttel O (2006) Monitoring of an infiltration experiment using the self-potential method. Water Resour Res 42(8):1–11. doi:10.1029/2005WR004840

Tchobanoglou G, Kreith F (2002) Solid Waste Handbook, 2nd edn. McGraw-Hill, New York

Tezkan B (1999) A review of environmental application of quasistationary electromagnetic techniques. Surv Geophys 20(3/4):279–308. doi:10.1023/A:1006669218545

Tezkan B, Goldman M, Greinwald S, Hördt A, Müller I, Neubauer FM, Zacher G (1996) A joint application of radiomagnetotellurics and transient electromagnetics to the investigation of a waste deposit in Cologne (Germany). J Appl Geophys 34(3):199–212. doi:10.1016/0926-9851(95)00016-X

Tezkan B, Hördt A, Gobashy M (2000) Two-dimensional radiomagnetotelluric investigation of industrial and domestic waste sites in Germany. J Appl Geophys 44(2–3):237–256. doi:10.1016/S0926-9851(99)00014-2

Tsourlos P, Vargemezis GN, Fikos I, Tsokas GN (2014) DC geoelectrical methods applied to landfill investigation: case studies from Greece. First Break 32(8):81–89

Valois R, Galibert P-Y, Guerin R, Plagnes V (2016) Application of combined time-lapse seismic refraction and electrical resistivity tomography to the analysis of infiltration and dissolution processes in the epikarst of the Causse du Larzac (France). Near Surf Geophys 14(1):13–22. doi:10.3997/1873-0604.2015052

Vargemezis G, Tsourlos P, Giannopoulos A, Trilyrakis P (2015) 3D electrical resistivity tomography technique for the investigation of a construction and demolition waste landfill site. Stud Geophys Geod 59(3):461–476. doi:10.1007/s11200-014-0146-5

Villain L, Sundström N, Perttu N, Alakangas L, Öhlander B (2011) Geophysical investigations to identify groundwater pathways at a small open-pit copper mine reclaimed by backfilling and sealing, pp 71–76

Walther EG, Pitchford AM, Olhoeft GR (1986) A strategy for detecting subsurface organic contaminants. The petroleum hydrocarbons and organic chemicals in ground water, prevention, detection and restoration. National Water Well Association, Houston, TX, pp 357–381

Wang T-P et al (2015) Applying FDEM, ERT and GPR at a site with soil contamination: a case study. J Appl Geophys 121:21–30. doi:10.1016/j.jappgeo.2015.07.005

Weigel M (1989) Self-potential surveys on waste dumps theory and practice. Detection of subsurface flow phenomena. Springer, Berlin, pp 109–120

Weller A, Frangos W, Seichter M (2000) Three-dimensional inversion of induced polarization data from simulated waste. J Appl Geophys 44(2–3):67–83. doi:10.1016/S0926-9851(00)00007-0

Whiteley RJ, Jewell C (1992) Geophysical techniques in contaminated lands assessment: do they deliver? Explor Geophys 23:557–565

Wijesekara HR, De Silva SN, Wijesundara DTDS, Basnayake BFA, Vithanage MS (2015) Leachate plume delineation and lithologic profiling using surface resistivity in an open municipal solid waste dumpsite, Sri Lanka. Environ Technol 36(23):2936–2943. doi:10.1080/09593330.2014.963697

Wijewardana YNS, Galagedara LW, Mowjood MIM, Kawamoto K (2015) Assessment of iInorganic pollutant contamination in groundwater using ground penetrating radar (GPR). Trop Agric Res 26(4):700. doi:10.4038/tar.v26i4.8132

Williams JHW, Lapham WW, Barringer TH (1993) Application of electromagnetic logging to contamination investigations in Glacial Sand-and-Gravel Aquifers. Ground Water Monit Remediat 13(3):129–138. doi:10.1111/j.1745-6592.1993.tb00082.x

Yin K, Tong HH, Noh O, Wang J-Y, Giannis A (2015) Mapping refuse profile in Singapore old dumping ground through electrical resistivity, s-wave velocity and geotechnical monitoring. Bull Environ Contam Toxicol 94(3):275–281. doi:10.1007/s00128-014-1427-y

Zacher G, Tezkan B, Neubauer FM, Hoerdt A, Mueller I (1996) Radio magnetotellurics: a powerful tool for waste-site exploration. Eur J Environ Eng Geophys 1:135–159

Zekkos DP (2005) Evaluation of static and dynamic properties of municipal solid-waste. University of California at Berkeley

Uranium Resource Development and Sustainability—Indian Case Study

A.K. Sarangi

Abstract Uranium, a highly concentrated source of energy is the basic fuel for nuclear power. Its mining, production, and use in generation of nuclear power, especially in a populous country like India helps in protecting the Earth from irreversible environmental damage—a giant step towards a sustainable development process. India's nuclear programme is based on unique sequential three stages—called "closed fuel cycle," aiming for optimal utilization of the indigenous atomic mineral resource (modest uranium and abundant thorium). The closed fuel cycle where the spent fuel of one stage is reprocessed to produce fuel for the next stage multiplies the energy potential of the fuel and therefore acknowledged as sustainable development compliant. Low grade narrow vein uranium deposits of India do not lend themselves for development on plain commercial considerations. The integrated economic model of nuclear power production programme of the country facilitates the uranium mining sector to adopt higher level of safety standards and environmental measures through application of technology implementing sustainable global practices. The technologies for uranium ore mining in India are appropriately chosen with an aim to achieve minimum disturbance to the topography of the area, minimize the generation of waste rock, use of the waste rock in underground, reuse and recycle of the liquid waste, continuous restoration of the voids created by underground mining, use of electro-hydraulic underground equipment in place of diesel powered ones, etc. These help in maintaining the operations within the absorptive capacity of local sinks for wastes. They contribute significantly towards achieving the goal of sustainable development. The uranium ore processing technologies in India are constantly upgraded with utmost consideration on maximizing the recovery, minimizing the discharge of effluents, and maximizing the recovery of by-products. Adopting a shorter processing route, measures to maximize the reuse of water, producing environmentally benign product, etc., are some of the distinctive features exemplifying values of sustainability. Management of uranium mill tailings is a matter of concern all over the world. In India, the technology for management of tailings has been constantly upgraded in

A.K. Sarangi (✉)
Uranium Corporation of India Ltd, Jamshedpur, India
e-mail: aksarangi@gmail.com

D. Sengupta and S. Agrahari (eds.), *Modelling Trends in Solid and Hazardous Waste Management*, DOI 10.1007/978-981-10-2410-8_6

line with the international practices. Tailing ponds have been designed with improved floor lining to prevent downward movement of effluent and monitoring mechanism to maintain the permissible discharge quality of water. Eco-restoration of the filled tailings pond is carried out with appropriate thick layer of soil which arrests radon emanation. Specified varieties of nonedible grass are grown which control soil erosion and do not cause radioactive incursion into the food chain through grazing animals. Mining industry, in general provides scope for growth of secondary industries in the neighborhood. Uranium industry in India has always been in the forefront in implementation of new technology in underground mining. Sustainability of these new technologies has been made possible by local entrepreneurs who progressively develop the competency to support need of critical components enhancing the skill base of the community. A skilled and trained society is thus created around the uranium mining centers of the country. Support for skill development and education to students in the area around uranium projects is considered very vital for integrated development programme of Indian uranium sector. Such programmes are initiated before the start of mining which helps in avoiding large-scale influx of trained manpower from other parts of the country. Displacement and disruption of settlements around mining sites is generally seen as a major cause of resentment in the communities. Balancing the sustainability of the community around uranium production facilities of India through appropriate mix of technology, environmental measures, social harmony, finance and governance is the hallmark of the growth of the sector.

Keywords Uranium · Energy source · Closed fuel cycle · Low grade uranium deposits · Mining · Uranium tailings management · Skill development · Sustainability

Introduction

Mankind has been pursuing the concept of sustainable development for a long time. Thomas Robert Malthus during 1700s thought about the dangers of population and coined the well-known statement "The power of population is indefinitely greater than the power in the Earth to produce subsistence for man" (Mark 2003). He had advocated for the preference of longer term strength of the economy to short-term convenience and emphasized on moral constraints as the best means of easing poverty—a thought similar to the theory of sustainability as is implicit today. Many ideas and thoughts have emerged since then striking various facets of mankind towards continued endurance. Gradually, these concepts took shape while debate raged on man's preoccupation with economic development at the cost of environmental concern and long-term maintenance of well being. Slowly, social aspects also found a place of eminence in the consideration widening and complicating the scope of debate. However, a consensus emerged in 1987 when Brundtland Commission articulated Sustainable Development as "Economic and social development that meets the needs of the current generation without undermining the ability of future generations to meet their own needs" (Our Common future 1987). In reality, sustainability refers to three broad themes; economic, social, and environmental that must all be harmonized and addressed to establish long-term viability of our community and the planet.

There have been significant changes in the world over the past two decades and the paradigm shift in technology and sciences itself has become one of the main drivers of economic and social development. Technological innovation is considered as the most important component of long-term economic growth and is a support structure for all strategies and policies aimed at ensuring sustainable economic development. Advancement in energy system driven by new discoveries and innovations has made a massive impact since the eighteenth century. Carbon dioxide emissions affecting climate and contributing to global warming is one of the major outcomes of the global energy consumption. As global energy demand continues to rise, reducing the energy sector's climate and environmental impacts remains imperative. Advancements in the energy sector are closely related to Research and development (R&D), which will be the foundation for a sustainable energy system.

The relationship between exploitation of various sources of energy and the development of civilization, particularly during the industrial revolution of the nineteenth century, validate the relationship between energy and modern economic development. With the rapid development in modern economy and process of industrialization, there exists a mass-scale use of fossil energies and dating back to the nineteenth century, the trend in annual global emissions of CO_2 shows a significant rise as shown in Fig. 1.

Energy plays a vital role in the growth and functioning of the world's economy. Dependence on energy will continue to rise as world population increases and standards of living progress. Historically it has been seen that increased energy use and mechanization to support ever-growing industrialization brings with it its own burdens such as environmental issues, health, safety, lifestyle, etc. Ideally, a matured society should find ways to keep a balance between what is socially desirable, economically viable, and ecologically sustainable through an adaptive process of integration.

Sustainable energy supply is an essential factor for sustainable development. However, the challenge lies in finding ways to reconcile this necessity and demand for energy resource with acceptable impact on the environment and the global natural resource base.

Energy resource is a prerequisite towards a sustainable life which in turn hinges on sustainable development. The energy sectors globally are defined by definite sources of energy available to them, viz., nonrenewable and renewable sources. The dependence on nonrenewable sources to meet the demand however is finite. Gradual replacement of fossil fuels requires consideration of many energy technologies and their deployment in specific applications. Currently, nuclear fission technology is the only developed energy source that is capable of delivering the demand of energy that will be needed to run modern industrial civilization safely,

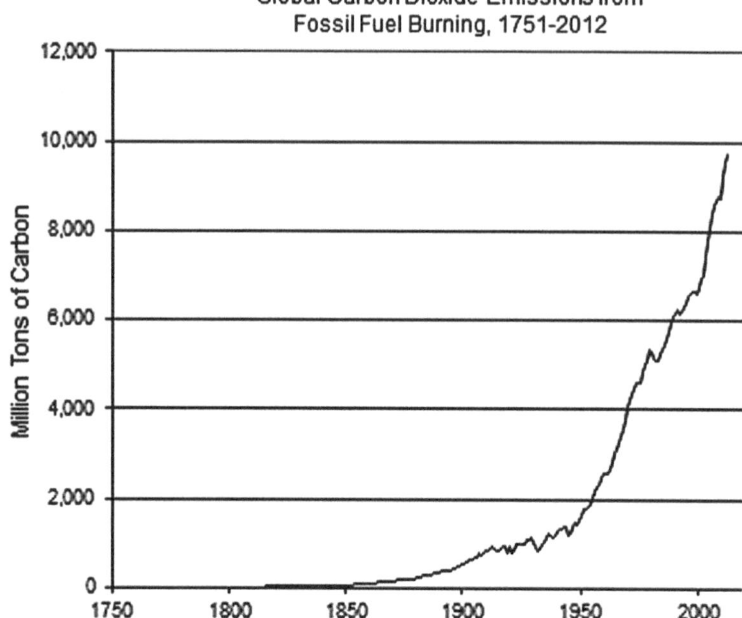

Fig. 1 Global carbon dioxide emissions over the years (*Source* www.climatechangeconnection.org/)

efficiently, consistently, and sustainably keeping in consideration the environmental issues and available resource base. As a result, nuclear fission has to play a major role in the revolution of the twenty-first century energy-supply system (Rynjah 2015). The process of nuclear fission of uranium produces heat. This heat is used in a nuclear reactor to generate steam, which drives a generator to produce electricity. Today, nuclear power plants provide 11 % of the world's electricity with US and France accounting for more than 50 % of the generated capacity. Presently, 31 countries host over 442 commercial nuclear power reactors with a total installed capacity of over 383.536 GWe and 66 reactors of 65.028 GWe are under construction (www.iaea.org/). Nuclear power has gained wide acceptance globally for its contribution to world energy requirements at low cost with lowest greenhouse gas emissions compared to any other electricity generation method. However, it requires meticulous research and robust technology for its production and strict control measures to prevent unwanted mishap.

Nuclear fission is an extremely potent source of energy and uranium is the basic raw material in this mission. Natural uranium occurs in most of the rocks in concentrations of 2–4 parts per million. Uranium is produced by mining and processing of uranium ore in much the same manner as other minerals like copper, gold, etc. The relative potential of uranium atom to generate electricity is much higher because of its high energy density shown in Table 1. Plutonium (^{239}Pu) produced

Table 1 Energy density by various sources of energy (*Source* http://oilprice.com/Alternative-Energy/Nuclear-Power/The-Inevitability-of-the-Nuclear-Age.html)

Material	Energy density (MJ/kg)
Wood	10
Ethanol	26.8
Coal	32.5
Crude oil	41.9
Diesel	48.8
Natural gas	55.6
Natural uranium	5,70,000
Reactor-grade uranium	37,00,000

by the transmutation of ^{238}U (a significant part of natural uranium) and ^{233}U produced by transmutation of ^{232}Th are also sources of fission to generate high energy with potential to generate electricity.

The present study shows that total identified world uranium resources have increased by 12.5 % since 2008 and they are sufficient for over 100 years of electricity generation based on the current requirements (World Energy Resources 2013 Survey). About 85 % of world uranium requirement for nuclear power plants is met by production from 19 countries and the balance requirement is met from secondary sources supply, notably inventories held by utilities, ex-military material, reprocessing, etc. (www.world-nuclear.org/). As the supply from ex-military material has dwindled during last 3 years with expiry of global agreements (Megaton to Megawatt programme, 1993 and US-Russian plutonium management and disposition agreement 2000), production of uranium from the primary sources is on rise. Kazakhstan, Canada, and Australia are the major producers of uranium and United States, France, and Japan are the major consumers.

Indian Energy Scenario and Sustainability

India, with a population of more than one billion, is the second most populous country in the world. The country has been attempting for an impressive growth of Gross Domestic Product (GDP) with improved quality life for the population and energy is an essential input for this development. India's energy policy till the end of the 1980s was mainly based on the availability of indigenous resources, coal being the largest source of energy. However, the energy policy of the country has been changing over time with an aim that a secure and robust energy supply is maintained at a reasonable cost and is generated from a variety of sources through diversified technologies. This is known as the '*energy mix*' which calls for selecting an optimal mix keeping in view the use of all resources and technologies. However, meeting the challenge of quantitative demand of clean, secure, and affordable electricity is the thrust area for India's endeavor towards sustainable development process and energy independence.

With this increasing need of energy, it has been felt that a balance mix of hydel, coal, and nuclear power is necessary for meeting the long-term requirement. Of all energy resources available, the known coal reserve of the country is estimated to last till the end of this century as a large amount of coal is required to be burnt for generation of heat in a thermal power plant. Coal-based power stations also pose serious environmental problems like generation of ash and emission of greenhouse gases and acid gases. Moreover, the reserve of thermal power grade coal is limited to only the eastern part of our country and transportation of large amount of coal to different corners shall demand a dedicated transportation network. A substantial part of coal resource of the country is amenable to underground mining and some part falls under environmentally sensitive areas. Factoring the commercial viability of underground coal mining in India, uncertainty over mining the part of the resource in environmentally sensitive area and standard mining losses, the estimated coal reserve is substantially less than the assumed resource.

The natural oil reserve of our country is inadequate and therefore has very limited scope to meet the long-term energy requirement. About 70 % of the domestic oil requirement is met by import. Uncertainty in supply of oil and its large price swing in the global market due to unpredictable geopolitical events have the potential to affect the domestic economy. Wherever possible, natural oil use needs to be conserved for special purposes, such as transport and the petrochemical industry. Similarly, the natural gas reserve is also very limited and therefore too valuable for uses in large-scale electricity generation.

Tapping the potential of hydro-power is limited to geographically suitable sites and entirely depends on good rainfall year after year. This also poses serious effect on ecology and social problems related to the displacement of vast population. Wind power and solar power sources have very limited scope to make any noticeable contribution to the energy security of any country (Sarangi 2015).

Nuclear power is the cleanest form of mass energy generation, producing no greenhouse gases like carbon dioxide (CO_2), sulfur dioxide (SO_2), and ash. Therefore, the growth of nuclear energy in developing and populous countries like India is a matter of great benefit for mankind in view of its potential to meet the energy demand as well as protect the Earth from irreversible environmental damage. Uranium being a highly concentrated source of energy is easily and cheaply transportable. In addition, the fuel cost contribution to the overall cost of electricity produced is relatively small. Therefore, persistent generation of abundant, affordable, and clean energy, i.e., nuclear power is a giant step towards a sustainable development process.

On the resource base towards nuclear power generation and geopolitical rationale, India is in a unique situation with moderate uranium resource and abundant thorium resource. Being a non-signatory to Non-proliferation Treaty (NPT), uranium and other nuclear material trading with India is subjected to strict international policy framework and compliance. Under such circumstances, India's nuclear power programme is oriented towards harnessing the near-infinite potential of thorium towards sustainable energy independence of the country. The relative abundance of different sources of energy in India is shown in Fig. 2.

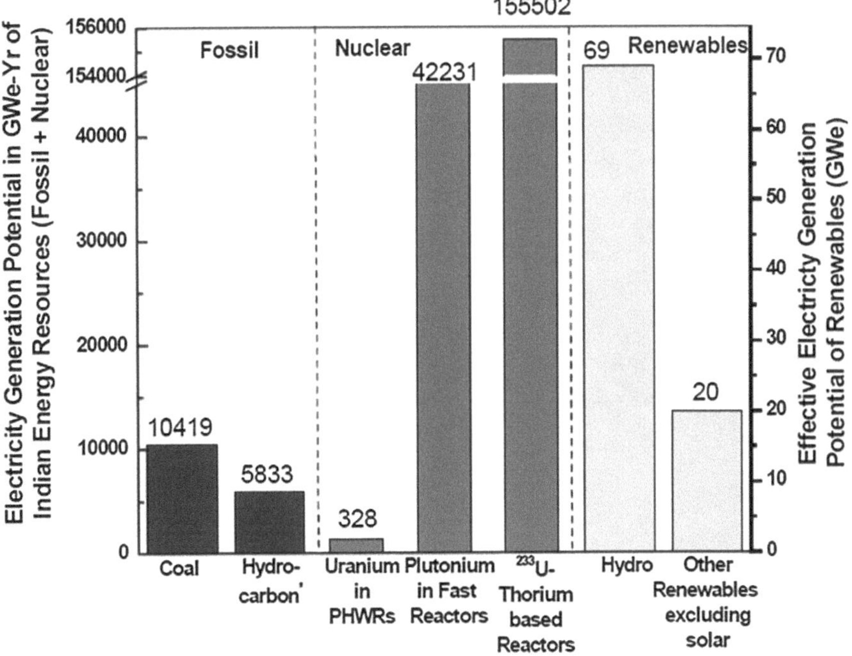

Fig. 2 Current India energy resources (*Source* Kakodkar 2008)

Most part of India's limited uranium resources are confined to five geological basins—Cuddapah basin (Andhra Pradesh and Telengana), Singhbhum Shear Zone (Jharkhand), Mahadek basin (Meghalaya), Delhi Supergroup of rocks (Rajasthan), and Bhima basin (Karnataka), whereas vast thorium resources are found along the large coastal belt of Odisha, Andhra Pradesh, Tamil Nadu, and Kerala. Some of the prominent thorium rich areas of the country are Chhatrapur-Gopalpur in Odisha, Bhavanapadu-Kalingapatnam-Bhimunipatnam in Andhra Pradesh, Manavalakurichi, besides Teri inland placers, in Tamil Nadu, Chavara in Kerala and Ratnagiri in Maharashtra, shown in Fig. 3.

As thorium is not a fissionable (but fertile) material, its industrial use is strategically planned in unique three-Stage nuclear power programme being pursued since the beginning of atomic energy programme in the country. Extensive indigenous capabilities have been developed in all aspects of nuclear power and associated fuel cycles. The large R&D base, skilled human resource, and facilities for continual advancement of human resource, industrial potential as well as robust regulatory framework, together forms an establishment of allegiance towards the objectives of sustainable development.

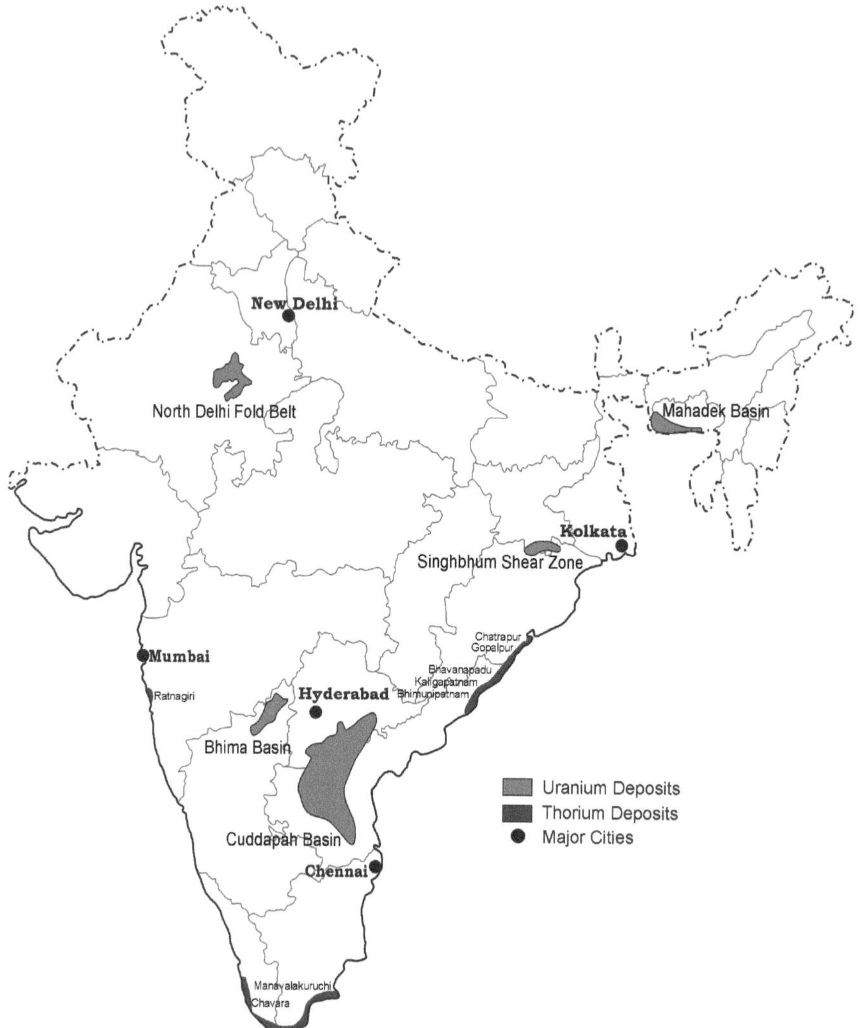

Fig. 3 India's uranium and thorium deposits

India's Three-Stage Nuclear Programme and Sustainability

India's nuclear power programme made a steady beginning with the formation of Atomic energy Commission in 1948 under the dynamic leadership of Dr. Homi J. Bhabha. The establishment of the Department of Atomic Energy in 1954 marked India's commitment towards research and development and peaceful uses of atomic energy. India, over the years has been able to acquire expertise on the every aspects of the power programme—the complete fuel cycle from uranium exploration, mining, fuel fabrication and electricity generation, reprocessing and waste management—a rare accomplishment in the world. India's nuclear programme is based on unique sequential three stages—called "Closed fuel cycle" designed by Dr. Bhabha who saw the potential of harnessing nuclear power in improving the quality of life of the millions of people by aiming to make optimal utilization of the domestic atomic mineral resource (modest uranium and abundant thorium). The closed fuel cycle where the spent fuel of one stage is reprocessed to produce fuel for the next stage, multiplies the energy potential of the fuel and therefore acknowledged as sustainable development compliant. It builds a base for harnessing the energy of all three fissionable materials ^{235}U, ^{239}Pu, and ^{233}U (Jain 2010).

STAGE 1-Pressurized Heavy Water Reactor (PHWR)
The reactors of this stage are fuelled by natural uranium which contains 0.7 % of ^{235}U (the only naturally occurring fissionable material) and remaining ^{238}U. ^{235}U undergoes fission to produce heat energy and ^{238}U undergoes transmutation to produce plutonium which is a fissionable material. Use of heavy water in the reactors (PHWR) increases the availability of neutron to convert ^{238}U to ^{239}Pu. The spent fuel from these reactors contains plutonium and depleted uranium which are recovered through reprocessing and used as fuel for second stage of reactors—a step towards sustainable energy supply.

India achieved complete self-reliance in this technology and PHWR programme is in the industrial domain. It is the mainstay of the three-stage nuclear power programme.

STAGE 2-Fast Breeder Reactor (FBR)
India's second stage of nuclear power generation envisages the use of ^{239}Pu obtained from the first stage reactor operation. It serves as the main fissile material. The plant is fuelled by mixed oxide of ^{238}U and ^{239}Pu. In FBRs, ^{239}Pu undergoes fission generating energy and ^{238}U undergoes transmutation to produce more plutonium. Since these reactors consume fuel to produce energy and also produce fuel for more energy, they are called breeder reactors. In the advance stage of this programme, thorium (^{232}Th) will be used in the core of the FBRs as blanket which will undergo transmutation to produce ^{233}U—a fissile material and fuel for the third stage reactors.

FBRs, technically produce more fuel that they consume and can thus take the electricity generating capacity at a very high level—a significant step towards energy sustainability. In India, a Fast Breeder Test Reactor (FBTR) is already in operation since 1997 and one FBR of 500 MWe is in advanced stage of completion. It is now proposed to use thorium-based fuel, along with a small feed of plutonium-based fuel in Advanced Heavy Water Reactors (AHWRs) which is expected to shorten the period of reaching the stage of large-scale thorium utilization.

STAGE 3-Breeder Reactor

The third stage of India's Nuclear Power Generation programme shall use ^{233}U as fuel. ^{233}U is obtained from the nuclear transmutation of ^{232}Th used as a blanket in the ^{239}Pu fuelled FBR (second stage). Besides being fuelled by ^{233}U, these reactors will have a ^{232}Th blanket around the ^{233}U reactor core which will generate more ^{233}U as the reactor goes operational thus resulting in the production of more and more ^{233}U fuel from the ^{232}Th blanket. Thus these reactors are called breeder reactors and India's vast thorium deposits permit design and operation of ^{233}U fuelled breeder reactors. As more of the ^{233}U in the fuel core is consumed, more and more ^{233}U will be produced helping to sustain the long-term power generation fuel requirement. These ^{233}U/^{232}Th based breeder reactors are under development and would serve as the mainstay of the final thorium utilization stage of the Indian nuclear programme. The currently known Indian thorium reserves amount to 3,58,000 GWe-yr of electrical energy and can easily meet the energy requirements during the next century and beyond (www.barc.gov.in).

India's Atomic Energy Programme is a mission-oriented comprehensive programme with a long-term strategy. The objective of the Indian Nuclear Power Programme is self-reliance through utilization of both fertile and fissile components of domestic atomic mineral resources (uranium and thorium). It focuses on building up nuclear power capability and sustainability in light of the nuclear material trade restrictions. It is generally acknowledged that nuclear power offers advantages of scale, reliability, and environmental benefits compared to fossil or other renewable energy sources (Bhardwaj 2013).

Nuclear Power Programme Towards Social Sector

With the use of natural uranium as fuel in India's nuclear power programme, many spin-off technologies arising out of those activities have contributed immensely towards quality of life of the people. Production of radio isotopes from nuclear power plants and research reactors has various applications including agriculture, health care, food preservation, industry, and research. A variety of radio isotopes are being produced in the research reactors at Trombay, meeting a major part of the demand in the country. Radiation Medicine Centre (RMC), in Mumbai, has become the hub for the growth of nuclear medicine in the country. Food preservation by irradiation reduces the percentage of food losses and provides means for improving food hygiene and facilitates export. Applications of radiation technology to agriculture have resulted in the improved varieties of seeds, which are contributing directly to food security in the society. Radiation technology applications in industries comprises of a wide range of activities, including radiography, water hydrology, gamma scanning of process equipment, use of tracers to study sediment transport at ports and harbors, flow measurements, pigging of buried pipelines and water hydrology, etc. These peaceful applications of radio isotopes are a promising step contributing towards sustainability in living standard of the society (www.dae.nic.in/).

Uranium Mining in India and Sustainability

The Indian nuclear power programme is distinctly a sequential three-stage scheme comprising of many advanced technologies. The nuclear power programme is geared towards maximizing the utilization of small indigenous uranium resources and fuelling the natural uranium to the PHWR programme (first stage) paving way for ultimate utilization of vast thorium resources. The design is an excellent example of sustainable energy availability as it provides power today for an energy starved country like India while supporting development of an abundant alternate source, thorium for future generations. However, uranium mining is a vital step playing a crucial role in determining the volume and direction of this development process.

Uranium Resources of India and Model for Sustainable Assessment Philosophy

Uranium deposits on Earth belong to more than one age and are of different types being restricted to distinct epochs, exhibiting the time-bound phenomena. The resources of the world are also very unevenly distributed, both in space and time.

Only a small part of the land mass in the Indian subcontinent is assumed to be geologically favorable for hosting uranium deposits. Uranium exploration in India, spanning over 50 years has brought out the presence of uranium deposits of five major types (vein type, sandstone type, strata-bound type, fracture controlled type, and unconformity proximal type) in different geological settings. Of the total uranium resource identified in India, Cuddapah basin in Andhra Pradesh and Telangana accounts for 52 %, Singhbhum Shear Zone in Jharkhand 29 %, Mahadek basin in Meghalaya 10 % and remaining in other states (shown in Fig. 4). However, uranium deposits of India are generally small in tonnage and of low grade. It is expected that future investigations adapting novel exploration strategies along with improvements of techniques focused towards deep-seated uranium mineralization shall enhance the uranium resource base of the country.

Resources are estimated mainly on the basis of geological information, which, as a rule is not available for commercial mining operations. Several techno-economic factors like mining, processing, environmental, legal and social aspects, etc. (internationally known as "modifying factors") are taken into consideration for estimating a modified subset of resources—called reserves. The resources established

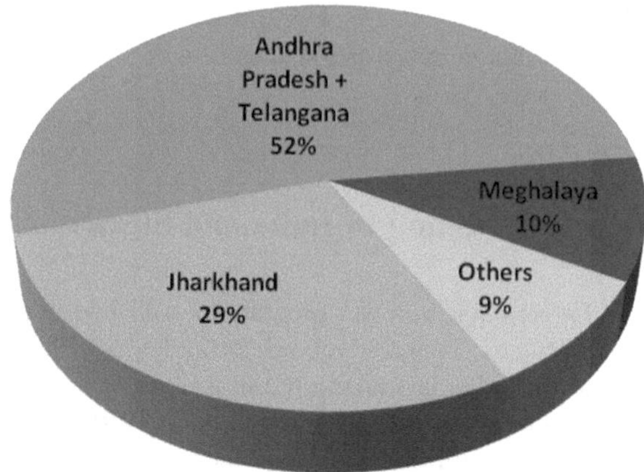

Fig. 4 Uranium resources distribution in India (*Source* Sarangi, Op. cit)

in some areas are sometimes not accessible for mining due to environmental considerations, adverse public perception on uranium mining, and sensitive administrative/political issues. As the above factors are of dynamic nature, the identified resources are reassessed with every change in above scenarios to estimate the reserves. Extensive pre-feasibility and feasibility studies prior to undertaking the mine development helps in understanding all techno-commercial parameters for sustainable operations.

Operating and Planned Uranium Production Centres in India

Uranium production in India to cater the indigenous need was launched with the formation of Uranium Corporation of India Ltd. (UCIL) in 1967 under Department of Atomic Energy. The operation started with the commissioning of an underground mine and ore processing plant at Jaduguda (1968) in Jharkhand (the then Bihar). Presently, activities (under operation and construction) of the country are confined to four different regions/zones (www.ucil.gov.in).

Singhbhum Shear Zone in the State of Jharkhand

The Singhbhum shear zone (SSZ) in eastern part of India hosts all the operating units (six underground mines, one opencast mine, and two ore processing plants) of the country. The underground mines at Jaduguda, Bhatin, Narwapahar, and Bagjata feed ore to the plant at Jaduguda. The plant at Turamdih is fed ore from Turamdih underground mine, Banduhurang open-pit mine, and Mohuldih underground mine.

Cuddapah Basin in the State of Andhra Pradesh

In the southwestern part of Cuddapah basin, low grade uranium mineralisation in siliceous-dolomitic-phosphatic-limestone hosted rock is extending over a length of 160 km. A large underground mine at Tummalapalle in this sector has been constructed and the ore is processed in the plant adjacent to the mine. The processing technology adopted for this plant is under stabilization. Plan for expansion of this mine and plant and construction of new mines and plants are being envisaged around Tummalapalle.

Uranium mineralisation in northern part of Cuddapah basin lies in the unconformity contact between arenaceous formation and basement granite. One opencast mine and three underground mines in this region at Lambapur-Peddagattu have been planned and the ore from these mines shall be treated in a mother plant at Seripally, 48 km from the mine site.

Bhima Basin in the State of Karnataka

High-grade mineralisation in limestone and granite hosted rock in this area is under exploratory mining at Gogi. An underground mine and plant has been planned at site.

Mahadek Sandstone Basin in the State of Meghalaya

The sedimentary basin, with a large areal extend hosts a large number of small to medium size uranium deposits in sandstone host rock in Meghalaya, N–E part of the country. Kylleng–Pyndengsohiong (former name Domiasiat) is one such large deposit where active mining and processing of uranium ore has been planned (Sarangi and Yadav 2011).

Uranium Mining and Milling Technologies and Sustainable Development Framework

Kazakhstan, Canada, and Australia contribute to more than 65 % of world's total uranium production. About 51 % of world uranium production comes from in situ leaching (ISL) method which has become increasingly important over last two decades because of its low cost advantage over conventional techniques of open-pit or underground mining and processing (www.neimagazine.com/). Other methods of uranium extraction are co-product or by-product recovery from copper, gold and phosphate operations, heap leaching, etc. Small amounts of uranium are also recovered from mine water treatment and environmental restoration activities.

Sustainable Evolution of Uranium Mining and Milling Technologies in India

Uranium mining in India during last five decades has evolved with adoption of appropriate technology in line with global developments without undermining the need for deploying of local skills, wherever required. The industry has undergone numerous phases of technological upgradation maintaining a balance among contemporary technological developments, incorporating scientific mining and processing with flexibility for upgradation, waste management (taking into consideration the carrying capacity of environment), understanding and coping with the socioeconomic needs and engagement with stakeholders along with involvement of local communities in decision-making. Mining processes have the potential to impact a diverse group of environmental and social entities. With improved planning, implementation of cleaner technologies and sound environmental management tools, performance in both the environmental and socioeconomic arenas will improve drastically and thus contribute enormously to sustainable development at the mine level. Transparency in communication, appropriately rehabilitating the project affected people, etc., is gloriously exemplified in the growth track of uranium industry of the country. This has assured sustainability at different stages of mining cycle—from exploration to mine closure.

Mining and Ore Extraction

India's first uranium mine at Jaduguda has undergone a transition in method of mining over the years from shrinkage stoping, open stoping to cut and fill method. Shrinkage stoping is applicable for steep dipping deposits with stable hangwall and footwall. Shrinkage and open stoping methods adopted in initial years were limited to shallow depth with low strata pressure. Although both the methods offered low manpower cost, significant loss of ore in pillars and broken rock within stopes (shrinkage stoping) and low productivity due to minimal mechanization called for adopting newer methods. In order to recover valuable uranium from the broken ore locked up in shrinkage stopes, Jaduguda mine had the distinction of recovering uranium by spraying and re-circulating the barren acidic ion-exchange solution into the stope and recovering the uranium from pregnant solution by precipitation—a novel method known as "in-stope leaching" (Gupta et al. 2003).

Currently, ore extraction at six underground uranium mines in Jharkhand is carried out by cut and fill method of stoping (Fig. 5). This method of mining permits the use of de-slimed mill tailings as backfill which contributes to sustainable waste management practices.

Cut and fill mining method permits retention of surface profile during the life of the mine and is thus restoration and reclamation friendly. This has also improved ore recovery and minimized dilution. Cut and fill mining method in Indian uranium industry has undergone rapid advancement facilitating use of alimak raise climber for raising, jumbo drill, pneumatic load-haul-dump (LHD), etc. Use of decline

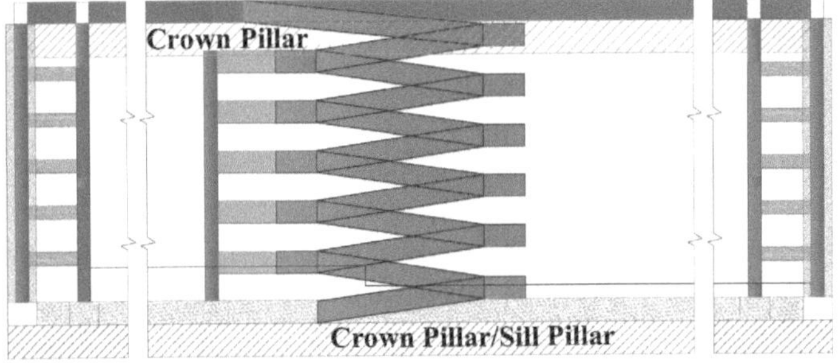

Fig. 5 Schematic layout of cut and fill stoping with ramp entry followed by UCIL

Fig. 6 Decline entry and mechanization in UCIL mines

method of mine entry in newer mines like Narwapahar, Turamdih, Bagjata, and Mohuldih has enabled use of trackless diesel LHDs which move directly into the stope to lift the broken rock (Fig. 6).

Cut and fill mining method has been practiced in most underground uranium mines in India in spite of its high cost. However, it helps in selective mining, amenability for mechanization, continuous restoration of voids through waste rock management, reuse and recycle of the liquid waste, etc. This method of mining allows extraction of ore from stopes with minimum disturbance to the surface and underground. It maximizes recovery through less dilution and utilizes the mined out space for waste rock disposal thus achieving the goal of sustainable development.

Ore Processing

Uranium ore is processed through hydro-metallurgical route in one of the two ways —acid leaching or alkali leaching depending on the ore characteristics. Both Jaduguda and Turamdih process plants in Singhbhum Shear Zone are based on acid leaching. The mined out ore undergoes crushing and grinding followed by thickening and re-pulping. The re-pulped material undergoes leaching in sulphuric acid medium under controlled conditions. The leached out slurry is then filtered and the liquor undergoes further purification through fixed bed ion-exchange in resin. The purified liquor is subjected to iron precipitation followed by uranium precipitation at different pH. Of all the possible precipitation schemes available for uranium, Uranium Peroxide is the most preferred product as it is purer and an environmentally superior product. The plant at Jaduguda produces Uranium Peroxide (yellow cake) which is further washed, filtered, dried, packed, and sent for further purification and fuel rod fabrication. Many advanced systems like Particle Size Monitoring (PSM) equipment in the grinding circuit, high rate thickener for thickening, belt filter in dewatering and filtration, and mechanically agitated leaching tanks, etc., have been successfully installed in Turamdih plant in Singhbhum (Fig. 7).

At Tummalapalle, ore processing is based on alkali leaching under high temperature and pressure. This technology is indigenously developed and being tried for the first time in the country (Suri 2010). A novel scheme of regeneration of reagents in the process plant is under implementation to operate the plant in a sustainable manner as the alkali leaching process is a complex and high cost technology.

Recovery of by-products ensures optimum utilization and sustainable extraction of resources. Valuable materials as by-products are also extracted at the process plants depending on the concentrations of these minerals in the feed. Jaduguda plant had undertaken recovery of copper, nickel, molybdenum, and magnetite from

Fig. 7 Mechanically agitated tanks and horizontal belt filter at Turamdih Plant

uranium ore. Magnetite recovery is being continued which not only adds value to the operation but also helps in conservation of valuable minerals. The plant at Tummalapalle shall produce appreciable quantity of sodium sulfate as by-product whose market potential is being explored. The copper ores of Singhbhum being mined in the vicinity of Jaduguda contain traces of uranium and Jaduguda plant has the unique distinction of recovering uranium from the tailings of copper concentrate.

Continuous monitoring of parameters and sampling at different stages of the processing activity ensures optimal utilization of raw materials with maximum efficiency. The process is designed in such a way to achieve high recovery and minimize loss of uranium in tailings. Recycling and reprocessing is done at different stages for sustainable consumption of raw materials and resources. This is vital on both recovery and environmental impact of waste discharge. Uranium recovery from tailings of nearby copper mines and recovery of by-products further contributes to the sustainable use of natural resources by uranium sector of the country.

Waste Management and Disposal of Tailings

Uranium resources in India are of low grade necessitating higher volume of extraction and processing of large quantity of ore resulting in generation of large volume waste in the form of waste rock from mines and mill tailings from processing plant. Waste management becomes a vital activity and waste management standards need to be constantly upgraded in line with the international practices and adopting new techniques, wherever applicable. Cut and fill mining method has the advantage of accommodating waste in the mined out stopes thereby reducing the amount of waste to be managed on the surface. The waste rock generated during ore production is sent back to the mined out stopes for backfilling. The remaining waste rock are spread in the dump yard in such a way to blend in with the natural terrain of the surroundings followed by plantation and revegetation process to maintain the aesthetics of the area (Fig. 8).

The coarse particles (sand fraction) generated from tailings of process plants are used as back fill in underground mines whereas the finer fraction of the neutralized tailings containing radio-nuclides and chemical toxins (which do not possess the qualities of backfilling material) are disposed in well-engineered impoundment facilities called tailings pond. Hills as natural barriers, improved floor lining, well-laid rain water drainage system, monitoring ponds in the downstream side, etc., are some of the critical considerations for a tailings impoundment facility in India (Acharya 2014). The decantation wells and channel ways are strategically placed at the inner periphery of the tailings pond allowing only the excess water to flow out through a well-laid drainage system to the effluent treatment plant.

Fig. 8 Reclaimed waste dump at Narwapahar

Maintenance of the storage or impoundment facility is also a significant step since tailings pond is an exposed facility and is sensitive to the surrounding environment. Reclamation of the tailings pond post maximum-limit-utilization is studied carefully to provide long-term stability of the tailings restricting the movement of contaminants. The reclamation scheme involves suitable soil capping, planting of specific grass, and long-term monitoring of all environmental matrices. Figure 9 shows a reclaimed tailings pond at Jaduguda.

Fig. 9 Reclaimed tailings pond at Jaduguda

Uranium Mining and Fiscal Model for Sustainable Development

Uranium which is considered as a strategic resource by most countries is subjected to a different kind of market force compared to other energy resources. It is associated with financial risks and future liabilities arising from nuclear activities. Indigenous nuclear power programme of India to a large extend has remained insulated from external market dynamics. Nuclear power plants are capital intensive but have low and stable marginal production costs with cost of uranium as fuel constituting only a few percent of the total cost of electricity.

Uranium mines under operation in India are of very low grade and small to medium size. Geological characteristics of the host rock and intricate ore geometry add to the complexities of the extraction process. Development of these deposits in merely commercial considerations is apparently unsuitable with any scope to undertake high level of safety, environmental measures, and application of advanced technology. However, in the Indian context the economics of operation in uranium mining sector and its upstream and downstream activities are integrated with the operation of nuclear power plants and generation of power. This model provides fiscal latitude to the uranium mining operations facilitating adoption of higher level of safety standards and environmental measures through application of technology, thereby implementing sustainable global practices.

The economics of power production and supply to the national grid at competitive price is the hallmark of nuclear sector in India unlike other countries where the uranium producers, fuel fabricators and nuclear plant operators look for favorable economics at every stage of operations. The concept of "ore-to-core" has provided the opportunity to implement the best practices at different stages in spite of adverse geological parameters.

Avoiding the transfer of large financial burden through time to future generations is a part of sustainable development goals. Future financial liabilities and costs associated with closure of uranium mines, decommissioning of process plants, and reclamation of tailings impoundment facilities are adequately set aside to ensure funds availability when needed in the future. This, in turn is largely included in the generation costs and passed on to current electricity customers.

Neighborhood Development Towards Sustainable Living

The human capital dimension of sustainable development consists of knowledge, education and employment opportunities, human welfare, equity and participation, and social capital in the form of effective institutions and voluntary associations, the rule of law and social cohesion. In this regard, nuclear energy is characterized by a net contribution to human and social capital. However, uranium mining industry which is in the front-end of nuclear fuel cycle is subjected to many challenges in terms of public acceptability and widely varying perceptions of the risks and benefits.

Uranium mining industry of India acknowledges that for securing a long-term and sustainable prospect, the activities must incorporate technical, business, and reputation assets to proceed in a cost-effective, environmentally acceptable, and socially supportive way. It is important to engage communities, governments, and investors for sustainable survival of the operations. While outlining the sustainable growth path, a greater emphasis on economic sustainability and community capacity building should be contemplated as a priority.

Uranium Corporation of India has taken a giant stride in this regard. The initiatives to transform the lives of the communities around the operations as part of its corporate rationale in the growth and development of the nation have been hailed by all stake holders. The Company engages experts for professional feedback through Need Assessment and Impact Study in the neighborhood. Based upon the detailed site study and findings, Primary Health, Primary Education, Vocational Training/Skill Up-gradation, Soft Skill Development, Self Employment Generation, Cultural and Sports initiatives and Environment and Mines Remediation schemes have been identified as the focus areas. These activities are implemented through national agenda for all corporate—Corporate Social Responsibility (CSR) initiatives. The key to the success of this process lies in the capability of the CSR teams to ensure open and honest two-way communication based on mutual trust, establish clear goals, effectively communicate the Company's role and place the community at the centre of the response.

One of its major sustainable development initiatives is anchored recently on "empowering women." In the very first year of its implementation, women of the local community of the nearby villages were reached out with an objective to nurture entrepreneurs whose earnings would contribute to family incomes through skills development programmes on Jari Saree Making, Stitching, Cutting & Tailoring, Leaf Bowl Making, Phenyl Making, Forklift Operator Training, etc. The Company has set up an Industrial Training Centre to impart basic skill training in areas like Fitter, Welder, Electrical, etc., to youths from the surrounding villages in order to make them self-reliant and job ready.

One of UCIL's key, CSR initiatives is an irrigation project in the nearby area taken up with a sheer motivation to transform the lives of 56 odd farmers and their families by facilitating them to be able to earn their livelihood and ascend to a better quality of life. These farmers were provided with irrigation pumps, free seeds, fertilizers, and manures along with periodical training and assured marketing of the products.

Upgrading the standard of living of nearby villages has played an important role in crystallizing community objectives. An enduring relationship with the neighboring villagers residing in the vicinity of the operating units through mutual respect, active partnership, and long-term commitments has been built over years, with an objective to inculcate amongst the local population the proud acceptance that they are the main stake holders and active participants in development endeavors of their society.

Conclusion

Uranium mining industry, all over the world is a victim of negative public perception because of legacy issues arising out of unscientific mining and processing activities causing contamination and environmental pollution during early years. With concerns over climate change, urgent need for power generation to meet the future needs, debate over potential of nuclear power to help mitigate against future greenhouse emissions and thrust for production of uranium in an environmentally and socially sustainable manner has gained momentum in many countries. With its high energy density, mining and transportation of substantially low volume of uranium counts high on sustainable development process.

India, in its unique nuclear power programme has accorded high priority to the use of all three fissionable isotopes—^{235}U, ^{239}Pu, and ^{233}U to meet the challenge of reaching energy independence through well-calibrated utilization of domestic uranium and thorium resources (Suri et al. 2014). The direction of this programme is influenced by indigenous atomic mineral resource base. It aims at ultimate goal of thorium utilization which is possible with optimal utilization uranium resource. Therefore, development of uranium deposits of the country is inevitable for the sustained supply of fuel to the indigenous nuclear power programme. Accordingly, the growth of uranium industry has shown an extraordinary up-trend during last one decade within a transparent regulatory framework. The industry is expected to expand further matching with the phenomenal growth of nuclear power generation in the coming years. UCIL has been pursuing this endeavor achieving technological and scientific excellence and aspiring to progressively achieve the capability of meeting the planned fuel requirement. A capability based on self-reliance has been acquired to take up newer challenges, as and when they arise, to meet current needs and to keep the growth rate intact.

Mining of low grade uranium reserve of the country with high productivity and with the approach to utilize low grade resources is a significant step towards sustainable development goals. Minimizing the process loss of uranium, maximizing the extraction of valuable by-products, reuse of process water towards progressive implantation of zero-discharge model, appropriate level of automation for monitoring process parameters, minimum discharge of tailings in environmentally acceptable form, etc., demonstrate sustainable production process in Indian uranium industry. All these have been made possible through the unique economic model of uranium extraction—integration with power generation.

Apart from supplying the raw material for nuclear fuel, the uranium mining industry of India has been contributing towards development of infrastructure, mining technology, and employment opportunity in the neighborhood. However, the key to gigantic growth in terms of volume and technological innovations in mining and ore processing sector lies in balancing four principal components of a sustainable society—producing companies, local communities, government agencies, and concerned environmental protection groups. The act of balancing as an endeavor should incorporate ample flexibility to maneuvre the direction of the journey towards meeting the needs of the present without depriving the future generations.

Acknowledgements The author wishes to acknowledge Chairman and Managing Director, Uranium Corporation of India Ltd. to give permission to publish the document.

References

Acharya D (2014) Overview of uranium production in India. In: Special ASSET Bulletin/Souvenir during 6th DAE-BRNS symposium on 'emerging trends in separation science and technology (SESTEC-2014) during February 25–28, 2014

Bhardwaj SA (2013) Indian nuclear power programme—past, present and future. Indian Acad Sci-Sadhan 38(5):781–784

Bhabha Atomic Research Centre (BARC) www.barc.gov.in/. Accessed 9th Nov 2015

Climate Change Connection www.climatechangeconnection.org/. Accessed 9th Nov 2015

Department of Atomic Energy (DAE) www.dae.nic.in/. Accessed 11th Nov 2015

Gupta R, Kundu AC, Sarangi AK (2003) Uranium mining, milling and tailings disposal—best practices. In: National seminar on environmental and sociological implications of mining (Coal, limestone and uranium) and exploitation of oil and natural gas in North–East India, North–East India Council for social Science, Shillong, pp 6–12

International Atomic Energy Agency (IAEA) https://www.iaea.org/PRIS/WorldStatistics/OperationalReactorsByCountry.aspx. Accessed 11th Feb 2016)

Jain SK (2010) Nuclear power—an alternative, p 3

Kakodkar A (2008) Evolving Indian nuclear programme—rationale and perspective. Indian Acad Sci Bangalore, July 2008, p 12

Mark P (2003) Introducing dynamic analysis using malthus's principle of population, p 2. http://wolfweb.unr.edu/homepage/pingle/Malthus.pdf

Nuclear Engineering International Magazine (NEIM) http://www.neimagazine.com/features/featuretreading-water-in-the-uranium-market-4684314/. Accessed 11th Feb 2016

Oil-price news http://oilprice.com/Alternative-Energy/Nuclear-Power/The-Inevitability-of-the-Nuclear-Age.html. Accessed 9th Nov 2015

Report by the United Nations Brundtland Commission (1987) Our common future. http://www.un-documents.net/our-common-future.pdf

Rynjah I (2015) Nuclear energy for sustainable development. Indian Mining Eng J 54(11):24

Sarangi AK, Yadav DS (2011) Uranium resources of india and potential for development. In: 17th convention of the Indian geological congress and international conference on new paradigms of exploration and sustainable mineral development: vision 2050 (NPESMD 2011), Nov 2011 at ISM Dhanbad, pp 99–110

Sarangi AK (2015) Nuclear power and status of uranium mining. Indian Geological Congress, pp 4–15

Suri AK (2010) Innovative process flowsheet for the recovery of uranium from Tummalapalle ore. BARC Newslett 317:6–7

Suri AK, Sreenivas T, Anand Rao K, Rajan KC, Srinivas K, Singh AK, Shenoy KT, Mishra T, Padmanabhan NPH, Ghosh SK (2014) Process development studies for the recovery of uranium and sodium sulphate from a lowgrade dolostone hosted strata bound type uranium ore deposit, Mineral Processing and Extractive Metallurgy

Uranium Corporation of India (UCIL) www.ucil.gov.in/. Accessed 9th Nov 2015

World Energy Council, World Energy Resources—2013 Survey, p 16

World Nuclear Association (WNA), www.world-nuclear.org/. Accessed 9th Nov 2015

Geohazard Modeling Using Remote Sensing and GIS

Sandeep Narayan Kundu

Geohazards

Geohazards are environmental conditions leading to widespread damage or risk to human life and property triggered by earth processes. These include a broad ensemble of natural and anthropogenic hazards that could cause disasters at the global scale, some examples of which are earthquakes, landslides, volcanic eruptions and floods. These catastrophic events disrupt the accessibility to and the availability of life sustaining commodities on which modern life rests on (Bostrum 2002; Smil 2008). Today, an improved understanding of the interactions between the hydrosphere, the atmosphere, the biosphere and the geosphere has led to the emergence and acceptance of climate change driven extreme events as geohazards. Rapid urbanization, deforestation, industrialization has impacted the biodiversity and the climate, feedbacks of which are resulting in an increased frequency and larger scales of geohazards. Present day events are dwarfed by the scale and magnitude of the ones which occurred in Holocene and should such a mega-hazard was to occur today, the consequences would be unparalleled.

Decadal statistics for the last century (Fig. 1) reveal that instances of drought, earthquakes, extreme temperatures, floods, landslides, wildfire and volcanic eruptions are on the rise. Economic losses from these events are also increasing exponentially with geographically large nations like the United States, China and India being the most vulnerable. Being sporadic, predictions of Geohazards are often challenging and therefore we tend to underestimate their frequency and intensity (Hempsell 2004; Wong 2014). Geohazards research is heavily reliant on indirect and incomplete historical evidences which make characterization and identification of the involved processes very difficult. Spatial statistics of past events and its direct impacts are therefore of paramount importance. It is however

S.N. Kundu (✉)
Department of Geography, National University of Singapore, Singapore, Singapore
e-mail: geosnk@nus.nus.edu.sg

D. Sengupta and S. Agrahari (eds.), *Modelling Trends in Solid and Hazardous Waste Management*, DOI 10.1007/978-981-10-2410-8_7

extremely difficult to collect and include information on the indirect impacts, which spiral out of immediate and direct ones on our environment and modern society, as they are difficult to assess and quantified with certainty. Recent events e.g. hurricanes Katrina (2005) and Sandy (2012), typhoon Haiyan (2013), the Indian Ocean tsunami (2004) and the 2011 Tohoku tsunami, have all illustrated that the risks associated with such events are difficult to estimate (Stein and Stein 2014) and that the procedures for reducing the disaster risk and mitigating the resulting losses are inadequate. Geohazards pose a big challenge for present day society as its impact cuts across geographical boundaries in today's increasingly complex built environment and global dependencies. A geohazard can therefore trigger global catastrophes that can disrupt the socioeconomic fabric of a region impacting its economy, food security and political stability. Floods and droughts are major threats that potentially could escalate to the whole planet through its spiraling impacts. With megacities and crucial industries situated in vulnerable areas, Geohazards like earthquakes, tsunamis, and volcanic eruptions might cause disasters that could exceed the capacity of the global economy to cope.

In the context of countries like India, where the geography and physical features are spread across varying topography, tectonic and climatic zones, the exposure to different Geohazards is high and is demonstrated through many events in the recent

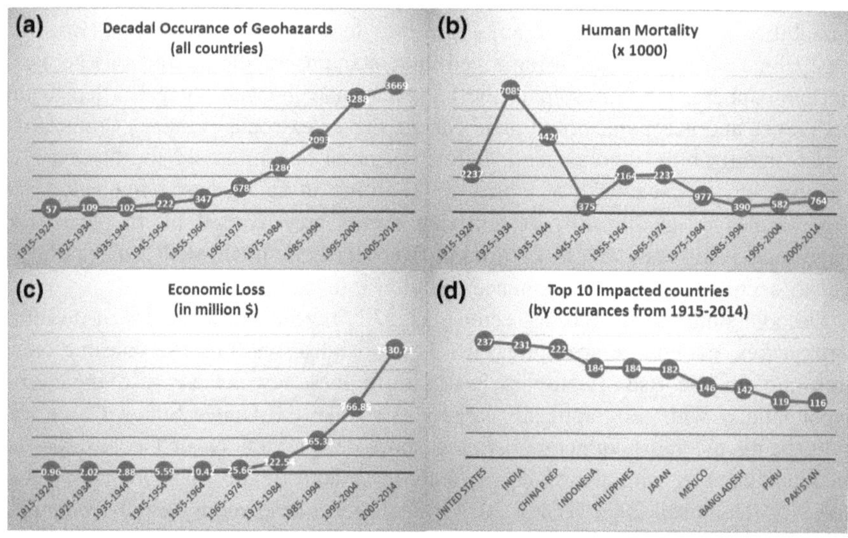

Fig. 1 Decadal statistics of geohazards. **a** Occurrences of geohazards. **b** Resulting human mortality. **c** Economic loss. **d** Top 10 countries by occurrence (1915–2014) (generated from data downloaded from www.emdat.de)

past. Floods in major rivers like the Brahmaputra, Ganga and Yamuna and tropical cyclones are becoming more prevalent than before. Intra-cratonic mega faults, along which major river flow and along which hydroelectric projects which fuel the emerging energy needs of the nation, pose seismic risks due to loading of water and sediments. In the Himalayan region, landslides and avalanches are easily triggered by seismicity or intense precipitation, have resulted regular loss of life and property. Like other parts of the world, geohazards are a serious threat to the socio-economic growth of India and with the government spending billions on rehabilitation of the victims; it becomes all the more important to include risk assessment for sustainable planning and development in these regions.

Earth observations through remote sensing from space, coupled with information from a ground-based network of sensors, play an essential role in increasing our resilience against natural hazards. They also provide policy makers with crucial information which help reduce the underlying disaster risk factors, making the society more adaptable. Insightful analysis using Geographical Information Systems (GIS) and timely dissemination of geospatial information on the risks and vulnerability helps the administration effectively mitigate the hazard. A systematic and structured approach using Remote Sensing and GIS applications can model the disaster and its hazard potential and identify vulnerable sections of the society beforehand which is helpful in mitigating the disaster and minimizing the losses.

Earth Observation and Remote Sensing

In the 50s, the launch of Sputnik and Explorer 1 by the Soviet and the US respectively, kicked off a space revolution that opened up a new way of observing the planet. Advances in space technology and computer processing led to the evolution of satellite observation systems spearheaded by several countries which led to scientific collaboration on earth observation at a global scale. The use of remote sensing for modelling geophysical phenomena over time led to better understanding of natural hazards. The knowledge gained from earth observation can now be seamlessly integrated with demographic and socio-economic data forming the foundation of risk mitigation planning and disaster response frameworks. As earth science oved towards studying the interactions between the hydrosphere, atmosphere, biosphere and geosphere as a complex system, a synoptic perspective as various scales and resolution was provided by remote sensing, making it an indispensable tool for modelling geohazards. Earth orbiting satellites complement observations from airborne or ground-based sensor networks for seismology, volcanology, geomorphology and hydrology. The spectral, spatial and temporal ranges provided by remote sensing platforms provide multiple perspectives of a region leading to estimation of several parameters which correlate well with ground based observations. Understanding the earth's dynamics and system complexity is now possible through integration of observations from multiple platforms improving our understanding on the earth's processes and feedbacks.

In the present day, satellite remote sensing provides a robust framework for continuous earth observation (Fig. 2), advancing scientific knowledge of our earth as a complex system of interactive processes, extremes of which are geohazards. Continuous observation provides temporal information that helps us identify trends which serve a precursor to geohazards and this leads to better preparation to face the hazards and even mitigate it. In the event of a geohazard, remote sensing helps us evaluate its impact through multi-temporal image analysis. Geospatial information products derived from earth observation and remote sensing feed the decision support systems we use today, which are being used at various scales of jurisdictions, forming the basis for comprehensive risk assessment and better-informed mitigation planning, disaster assessment and response prioritization.

Geohazards often occur in regions with restricted political and physical access. In such cases, remote sensing becomes the sole means for hazard assessment. Its robust data is amenable for analysis in geospatial systems that support image classification, digital terrain modelling, change analysis and integrated multivariate analysis incorporating ground based sensors. Global Positioning Systems (GPS), for example, provides space based geodetic measurements that are combines with

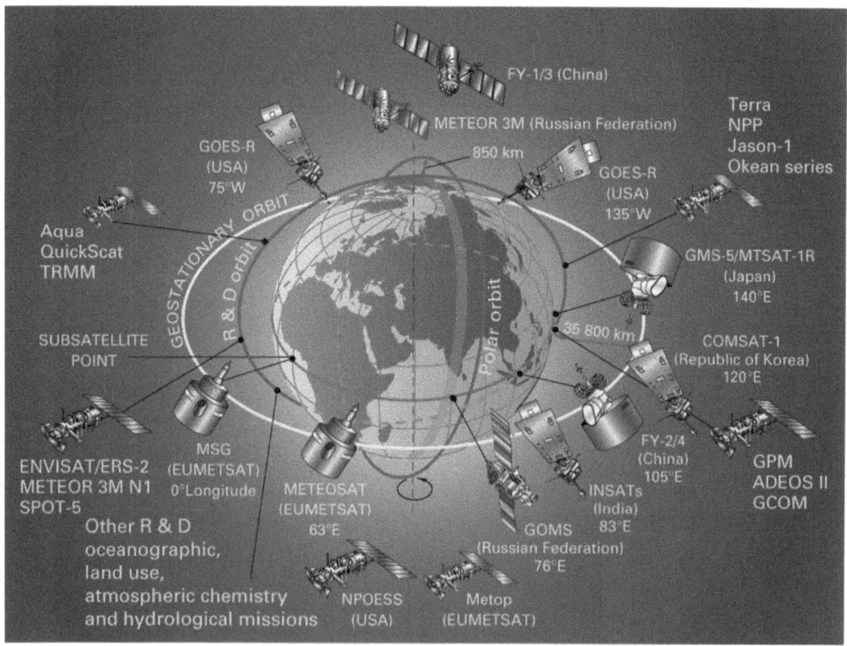

Fig. 2 The WMO Global Satellite Observing System, 2009 (from Mohr 2010)

ground seismic observations for modelling lithospheric movements leading to hazard and risk zonation maps for earthquakes. High resolution digital elevation models derived from Aster, Modis and Hyperion are contributing significantly to assessing and modelling hazards like landslides, avalanches and floods. Modern satellite based gravity field data (e.g. GRACE) are addressing short term total water variations on land and oceans, trends of which are used to predict drought. Coastline monitoring from space (IRS P4 Oceansat) provide measures for identifying risks of coastal flooding, erosion and their mitigation planning.

GIS Analysis

Geographical Information Systems (GIS) comprise tools for capturing, storing, analyzing, and presenting spatial data sets. GIS has evolved to handle large data sets from disparate sources and in evaluating complex spatial relationships between various measurements from ground and space borne sensors, which makes it an important tool for geohazard studies. GIS has been used for identifying precursors to, modelling the impact of and identifying causative factors leading to a geohazard. For example, the previous locations of events (earthquakes or landslides) can be used to study the spatial relationship between geological structures, moisture distribution, thermal conditions, and geomorphic features to identify the vulnerability of another region leading to the hazard. High-frequency Earth observation data can be analyzed to generate trends that not only can forecast events like blizzards or tropical cyclones but also can serve as an early warning to vulnerable community preparing them well in advance to face it.

GIS aids spatial analysis of pre and post event scenarios lead to quantification of the impact and its severity which is useful for authorities for their restoration work. Development of spatial risk model that facilitates visualization of what-if scenarios in the event of a geohazard occurring is now possible to model using modern GIS tools. Today, the role of GIS extends beyond spatio-statistical analyses into web-based applications and communication systems which now form the backbone of decision support systems in many countries.

Predictive Analysis

Geohazards do not occur regularly and are constrained by spatially spread environmental factors. Geospatial predictive analyses utilizes spatial correlation among these environmental constraints and by using and comparing historical data, the risk of that hazard can be modelled in an area. Geospatial predictive modeling is a process for analyzing events using a geographic filter in order to make statements of likelihood of occurrence of the hazard. This is achieved through deductive or inductive methods. The former involves subjective information for describing the relationship between the factors and the occurrences of the geohazard. The accuracy of the deductive model is limited by the depth of qualitative data inputs to the model. Inductive methods, on the other hand, rely on the empirical relationship between factors that is quantitatively defined. This makes the methods amenable for implementation in computing systems using automated and iterative algorithms, results of which lead to identification of precursors to geohazards and their integration into early warning systems.

Predictive analysis can be extended into what-if analysis where hypothetical scenarios can be modelled to inductively or deductively assess the extremes of the risk or vulnerability of the geohazard to a region. A dam, for example, not only increases the risk of reservoir induced seismicity and trigger earthquakes which has the potential to flood the downstream region, in case of the failure of the dam. These spiraling hazards can be modelled using GIS at the planning stage of the hydro-electric project and its environmental risk evaluated, based on which authorities can decide on advancing or scrapping of the project. Such analyses are finding importance in the policy making framework of progressive governments. What-if analyses are the basis for environmental impact assessment and are a means to assess the hazard to the environment for activities like large-scale construction, mining or oil exploration, or transportation of hazardous materials on land and on sea.

Change Analysis

Change analysis is essentially an image processing technique which detects the differences in a region before and after an event. It involves a pair of congruent remote sensing images, one acquired before the event and another after. Such analysis is important for assessment of the impact of a geohazard.

The approaches to detecting changes depend on the nature of the hazard and also on the characteristics of the satellite image. Methods are categorized into two main groups, depending on whether image classification is required before or after the changes are detected. The first group of change detection methods (Fig. 3), involves classification into categories followed by their comparison to detect changes. This is comparable to the deductive approach in predictive analysis and requires extensive human expertise in classifying the images. The second group involves image differencing or derivation of the change vector, result of which is then sliced into categories which bring forth the factors which determine the impacts of the event. This method is computationally easier to implement and the human subjectivity is only introduced when thresholds for isolation of the categories is concerned.

Applications

We shall now look into some examples of the application of GIS and remote sensing on modelling different geohazards. The occurrence of a disaster depends on two primary factors; the probability of occurrence of a geohazard and the degree of loss

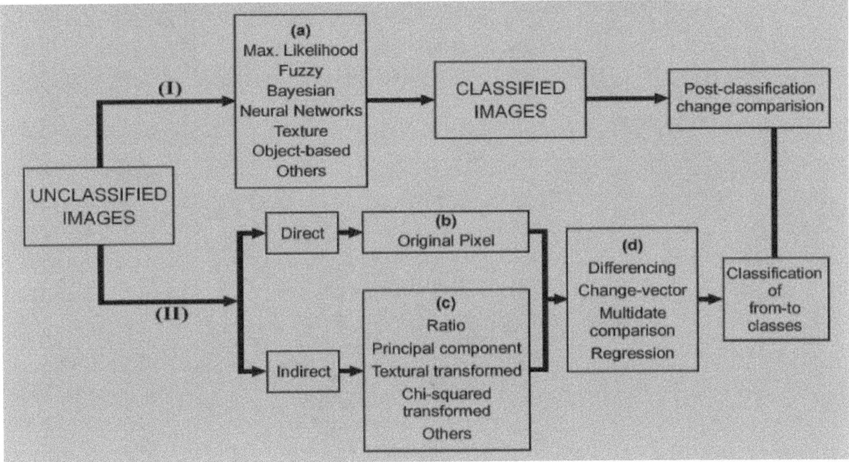

Fig. 3 GIS analysis framework for remote sensing data applicable for geohazard modeling (Siu and Lan 2008)

resulting from it. The former is the likelihood of the hazard itself, whereas the latter is the vulnerability. GIS utilizes remote sensing data to understand these factors through hazard and vulnerability analysis resulting in an actionable risk map.

Earthquake Hazards

An earthquake is the seismic shaking of the Earth's surface, resulting from the sudden release of energy from a stressed location in the earth's crust. Earthquakes have the potential to push back a city to ground zero with devastating impacts on life and the build environment (Fig. 4).

Satellite remote sensing systems not only offer spatially continuous information about the tectonic landscape, but also contribute to the understanding of specific fault systems. Combined with ground network data, remote sensing enables a better understanding of displacements, and validation of slip models that are cast in a regional setting of tectonic strain (Cakir et al. 2003) and help constrain source characterization. GIS is used to assess earthquake risk and hazard locations in relation to populations, property, and natural resources. The integration of spatial information on past occurrences, proximity to crustal plate boundaries, other structural geological elements and associated risks like landslides and floods are key for such geospatial analysis. Risk maps are deliberated at local, regional and national levels based on which land use and development activities are planned. For example, constructions in a high damage risk zone needs to incorporate an earthquake resilient design that can withstand the likelihood of a quake of expected intensity.

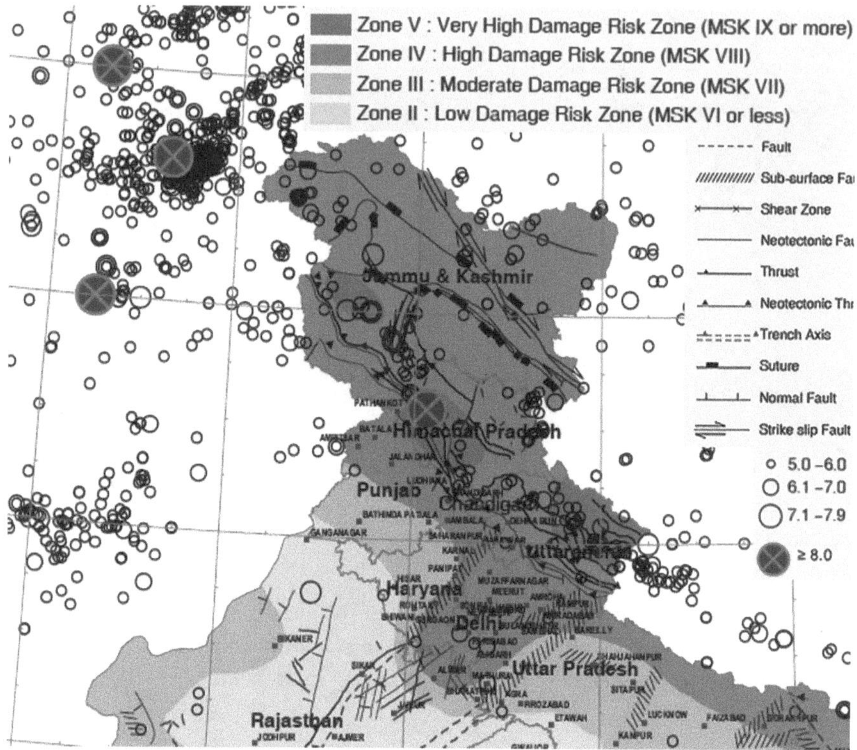

Fig. 4 Earthquake hazard map of Northern India showing risk zones calculated using past occurrences and regional geological structures (*Source* www.bmtpc.org)

Volcanic Hazards

Volcanoes erupt without warning leading to events like lava flows, lahars, ash falls, debris avalanches, and pyroclastic density currents. These have killed people and damaged several homes this century and pose a threat for the future. Modeling volcanic hazards require space and field-based observational data like the three-dimensional seismicity in the area; geodetic and gravimetric measurements of the deformation at the volcanic edifice that characterize the faults, fractures, landslides and flank instabilities, rift systems, and calderas; characterization of gas and ash emissions by composition and flux; and characterizing and monitoring of heat flux in the vicinity (Fig. 5).

Remote sensing defines a new paradigm for volcanological observations (Pieri and Abrams 2004) through spectroscopy in optical and thermal spectrums permitting the identification, isolation, and estimation of surface deposits and its composition, surface temperature, topography, and surface deformation aided by the Shuttle Radar Topography Mission (SRTM) provide invaluable topographic data in a temporal frequency (Steven et al. 2003). These datasets can be incorporated in GIS-based open

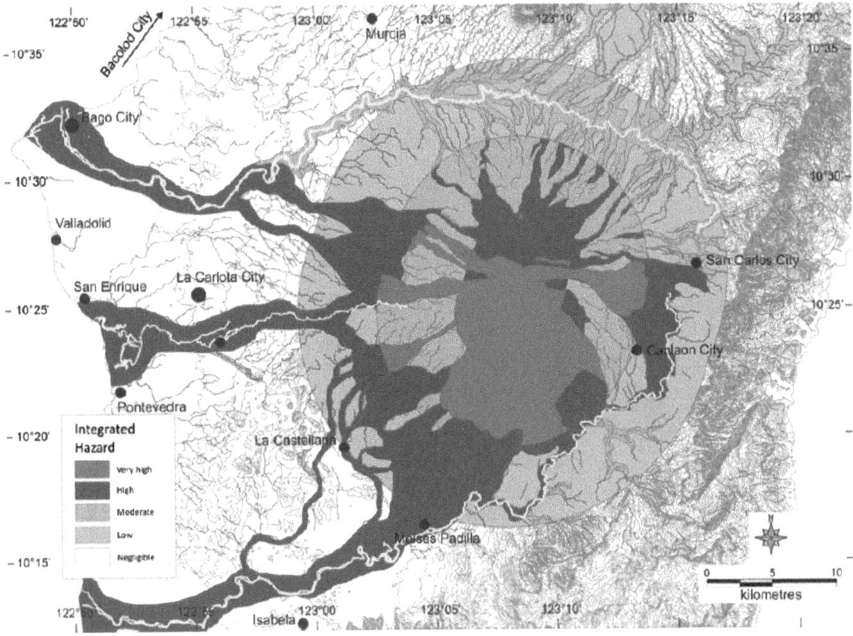

Fig. 5 Multi-hazard risk map in Kanlaon, Philippines (Neri et al. 2013)

source tools, e.g., VORIS (Felpeto et al. 2007) for volcanic hazard assessment, modeling, and generating risk maps.

Floods and Landslides

Floods are among the most frequent natural hazards in the world, claiming the largest amount of lives and property damage. Remotely sensing data analysis supports reconstruction of past erosion, deformation, and quantification of tectonic, climatic, biological, and anthropogenic inputs to the evolving landscape that underpins the risk and impact assessment of such hazards.

Observational geospatial datasets acquired over time characterize the changing surface, subsurface and hydrology of an area that can be quantified and modeled to predict flood hazards. High-precision sub-meter digital elevation models (DEMs) characterize morphology of catchments that can incorporate meteorological data, simulate runoff, quantifying threshold limits exceeding which floods or landslides are bound to occur. Floods and landslides have several commonalities in their occurrences and pose an increased threat to open cast mining (Fig. 6). Modelling their triggers is critical to the safety of people and equipment dwelling in the region.

Damage from instances of floods and landslides can be assessed using change analysis of a variety of satellite data sets from SPOT, Landsat to the recent high-resolution WorldView. An example of such assessment is for the South

Fig. 6 A devastating landslide at the Bingham Canyon Mine outside of Salt Lake City, Utah (*Source* Daily Mail UK, 1 June 2013)

Carolina floods in 2015 (Fig. 7), where more than a dozen dams were breached, entire neighborhoods were swamped, hundreds of roads were made impassable and more than a dozen of people were killed.

Coastal Hazards

Eleven of the 15 largest cities in the world are situated near coasts. Coasts are vulnerable to geohazards like cyclones, tsunamis, and in the wake of rapid climate

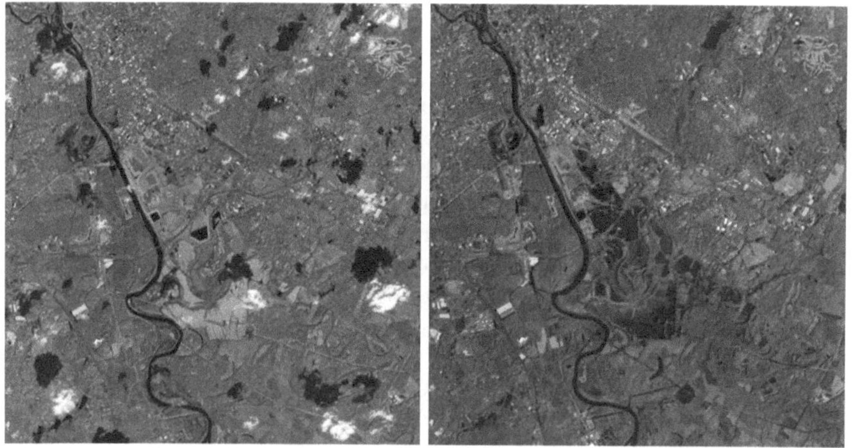

Fig. 7 True Color Composite of Landsat 8 OLI before (*left*) and after (*right*) the South Carolina floods. *Source* NASA Earth Observatory

change, sea level rise. The effects of coastal geohazards are spatially non-uniform due to local variables which involve interactions between lithology, geomorphology, climate, wave and currents (Gornitz 1991) and past storm frequencies.

Integrated SAR and Landsat TM imagery have been used to monitor changes in coastal geomorphology and land cover, flood and erosional damage, and to facilitate planning and maintenance of mitigations (Singhroy 1995) based on which GIS-based coastal flooding hazard zonation maps can be prepared, published, and disseminated to stakeholders.

The impact of coastal geohazards like cyclones can be studies using indices derived from earth observation satellites. Kundu et al. 2001 effectively demonstrated the use of IRS-P4 OCM to derive indices characterizing vegetation and coastal waters before and after the Orissa super cyclone of October 1999 using Image processing and GIS tools to access the coastal damage (Fig. 8). With the current day advances in satellite observation and geo-computing capability space measurements of ocean vector winds from SeaWinds on QuickSCAT and Meteosat can be combined with associated precipitation measurements to model the strength and damaging potential of coastal storms.

Natural Resource Exploration and Exploitation

Anthropogenic activity is increasingly playing an important role in triggering several geohazards. The seismic threat arising out from impounding of rivers for

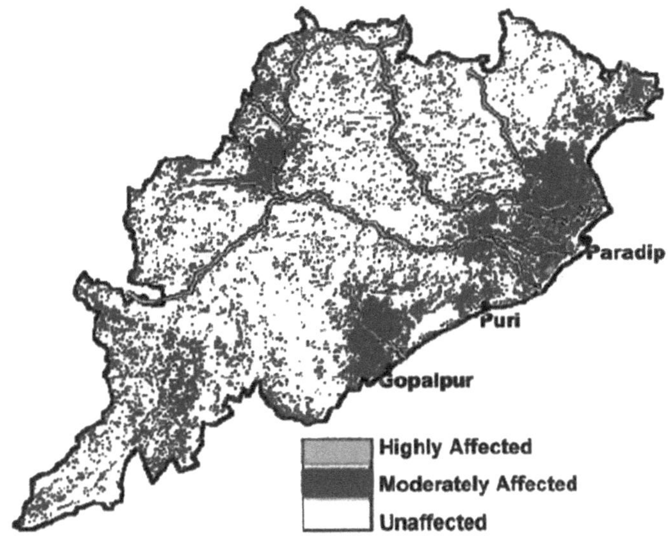

Fig. 8 NDVI difference image showing regions affected by the 1999 Orissa super cyclone, India (Kundu et al. 2001)

Fig. 9 NASA's Aster false color composite before (*left*) and after (*right*) the eruption of the mud volcano Lusi at Sdiaorjo, Indonesia, the biggest mud volcano of the world. The vent is suspected to have resulted from drilling-induced fracturing at the site (*Source* Wikipedia)

hydroelectricity generation, increases the chance for an earthquake which has the potential to spiral other geohazards like floods and landslides. Open cast mining methods for coal and copper influence changes in the topography, land cover and ground water levels, which can potentially cause landslides, erosion and air-water pollution, spiraling effects of which can be seen through generations. Drilling wells for oil and gas can encounter overpressured formations leading to blowouts, oil spills and mud volcanism. The Macondo disaster and the infamous Sdioarjo mud eruption in Indonesia (Fig. 9), both of which resulted from blowouts during drilling for oil and gas, are examples of such geohazards. Transportation of mined material also poses risks to the environment, so does excessive use of carbon-based energy resources that's driving global warming and climate change.

Opportunities and Challenges

The increasing availability of high-resolution (temporal, spatial, and spectral) satellite remote sensing platforms is supporting continuous earth observation and the evolution of efficient methods for rapid monitoring of land cover/land use and their changes. This improves chances of predicting a geohazard allowing timely decisions to be taken to assess risk assessment during extreme events (e.g., hurricanes, forest fires, earthquakes, floods, and epidemics) through spatial-statistical analyses and detection of trends. Currently there is a huge opportunity to improve understanding of each geohazard, not only in identifying the driving environmental factors, but also in establishing the empirical relationships that shall reduce subjectivity and improving confidence in the geohazard models and their subsequent integration in disaster management and decision support systems.

Despite the availability of the huge literature on land cover classification and change detection, we are still far from being able to automate these tasks via remote sensing and GIS. The high variability of ground conditions and integrated analysis of disparate geospatial datasets from various sensors operating at varying spatial and temporal precisions add another level of complexity to GIS-based predictive modeling and impact analysis. The quest for useful approaches and methods for rapid land cover identification and change detection remain a very challenging task and offer an opportunity for further research in geohazard modeling.

References

Bostrum N (2002) Existential risks—analyzing human extinction scenarios and related hazards. J Evol Technol 9. http://www.jetpress.org/volume9/risks.html

Cakir Z, Chabalier JB, Armijo R, Meyer B, Barka A, Peltzer G (2003) Coseismic and early post-seismic slip associated with the 1999 Izmit earthquake (Turkey), from SAR interferometry and tectonic field observations. Geophys J Int 155(1):93–110

Felpeto A, Martí J, Ortiz R (2007) Automatic GIS-based system for volcanic hazard assessment. J Volcanol Geotherm Res 166:106–116

Gornitz V (1991) Global coastal hazards from future sea level rise. Glob Planet Change 89(4):379–398

Hempsell CM (2004) The investigation of natural global catastrophes. J British Interplanetary Soc 57:2–13

Kundu SN, Sahoo AK, Mohapatra S, Singh RP (2001) Change analysis using IRS-P4 OCM data after the Orissa super cyclone. Int J Remote Sens 22(7):1383–1389

Mohr T (2010) The global satellite observing system: a success story. WMO Bull 59(1):7–11

Neri M, Le Cozannet G, Thierry P, Bignami C, Ruch J (2013) A method for multi-hazard mapping in poorly known volcanic areas: an example from Kanlaon (Philippines). Nat Hazards Earth Syst Sci 13:1929–1943

Pieri D, Abrams M (2004) ASTER watches the world's volcanoes: a new paradigm for volcanological observations from orbit. J Volcanol Geotherm Res 135(1–2):13–28

Siu N, Lam N (2008) Methodologies for mapping land cover/land use and its change. In Liang S (ed) Advances in land remote sensing. Springer Science, pp 341–367

Smil V (2008) Global catastrophes and trends: the next 50 years. MIT Press

Stevens NF, Manville V, Heron DW (2003) The sensitivity of a volcanic flow model to digital elevation model accuracy: experiments with digitized map contours and interferometric SAR at Ruapehu and Taranaki volcanoes, New Zealand. J Volcanol Geotherm Res 119(1–4):89–105

Wong I (2014) How big, how bad, how often: are extreme events accounted for in modern seismic hazard analyses? Nat Hazards 72(3):1299–1309

Singhroy V (1995) SAR integrated techniques for geohazard assessment, natural hazards: monitoring and assessment using remote sensing techniques. Adv Space Res 33(3):205–290

Stein S, Stein J (2014) Playing against nature: integrating science and economics to mitigate natural hazards in an uncertain world. Wiley Works, Wiley, Oxford

Modeling Trends in Solid and Hazardous Waste Management

The Change from Aerobic to Anaerobic Land Filling and the Increase of Correspondent Emissions and Expenses

Andre Luiz Bufoni

Abstract In this chapter, we discuss the main policy to waste management destination, considering the immediate sanitation concern in developing countries. From the analysis of literature and many waste management projects presented for United Nation registration as clean development mechanism (CDM), we draw some deductions. Because of the regulatory shortcomings and pitfalls; landfill sites configuration from aerobic unmanaged to anaerobic managed; poor administrative activity; and the increase of costs and the capital shortage, the change can lead to a greater overall GHG emission. Thus, the recent studies and the world reports about this issue indicate the regulatory environment as the first bottleneck to be aware of to alleviate the initiatives barriers.

Keywords Landfilling · GHG · CDM · Barriers

Contextualization

Since our consuming patterns are still unsustainable, the product life cycle is becoming increasingly shorter (especially electronic devices), recycling have limits, and prevention and reuse policies rates are very modest, the management of waste will also be for a long time a relevant and continuing global issue (Bartl 2011, 2014).

However, nowadays, while developed countries strategies focus on the resources conservation and landfilling diversion (Fig. 1), the developing ones strive to protect the health and wellbeing of people that still suffers due to inadequate waste man-

A.L. Bufoni (✉)
Federal University of Rio de Janeiro, Rio de Janeiro, Brazil
e-mail: bufoni@facc.ufri.br

D. Sengupta and S. Agrahari (eds.), *Modelling Trends in Solid and Hazardous Waste Management*, DOI 10.1007/978-981-10-2410-8_8

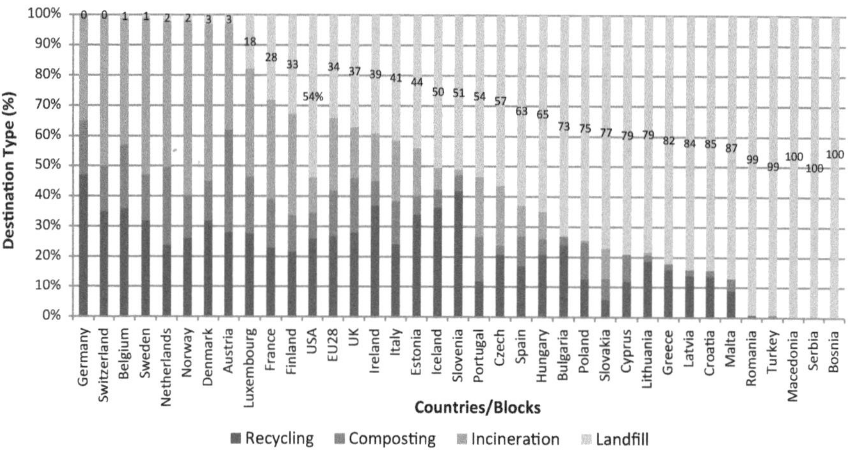

Fig. 1 Waste destination USA and EU (2012). *Source* Bufoni et al. (2014)

agement practice. For developing countries (DC) the protection of human health is still the number one objective (Brunner and Fellner 2007; Ciplak and Barton 2012; Bufoni et al. 2014). Certainly, and despite the fact that the waste management requirements are continually increasing for the entire world, for sure these countries face different cultural, technological, regulatory, and skilled labor barriers, but the financial constraints are common to all DC and pervades all others (Carvalho et al. 2011; Bufoni et al. 2015).

Per example, an investigation of large waste management projects of Clean Development Mechanisms (CDM) under Kyoto Protocol Instruments shows that the reality of the Municipal Solid Waste Management System in DC is that more than 70 % of the registered projects—in order to receive the respective tradable certified emission reductions (CER)—focus landfill gas treatment through flare or to generate electricity, usually using clean biogas (LFG) reciprocating engines (GENSET). However, the electricity generation usually is a second project stage and may not be implemented due to the LFG production uncertainties, and the high cost of the engines. The equipment pays it costs, what is not attractive to a regular financial investor (Bufoni et al. 2015).

Figure 1 shows the waste final destination of the European Union and the United States of America. The DC landfilling reality is more like the right part of the graph. However, this also means that the landfill profile is changing from an unmanaged swallow aerobic open dumps to intensive managed anaerobic site in most DC countries, what according to IPCC Guidelines for National Greenhouse Gas Inventories has its consequences, increasing costs and methane potential emissions. Nevertheless, the literature indicates that composting and anaerobic digestion (AD) are the two preferred municipal solid waste organic fraction (OFMSW) treatment option for achieving effective carbon reductions in developing countries (Barton et al. 2008; Rogger et al. 2011), while the (developed) world most significant waste directives also recommends the use of a new generation of Rankine cycle incinerators (HIRN) to recover the energy and to the landfilling volume reduction (European Comission 2008; EPA 2011).

In this chapter, developing countries environmental or distal context, like cultural and structural factors, are considered highly relevant to shape the final behavior from the intention of those tasked with the implementation of a good waste management system, where the municipality officials also have the management responsibility. However, it is not uncommon that these officials are political and regulatory dependents, and they lack the decision-making power (Godfrey et al. 2013). In conclusion, many developing countries are reestablishing strategies, reordering priorities, increasing standards, and extending regulations for the management of their waste prevailing practices and policies.

Consequently, the aim of this part is to present the circumstances supporting the developing countries "landfilling revolution", and critically discuss the adopted long-term strategy (if any) to manage the municipal solid waste. This work underpins this discussion of the changes in progress in four different dimensions: (1) the aim and the strategy of the national waste management policy (2) the profile of landfills sites, (3) the theoretical financial analysis on costs; and (4) the projected change in the emission of greenhouse gases on the model in IPCCC.

Contradictorily, waste management literature indicates that "waste management requires a more systems-oriented approach that addresses the root causes of the problems" (Singh et al. 2014), but a recent review points out that the aim of the studies rarely are strategic, or the barriers of the projects (Allesch and Brunner 2014), perhaps 'en passant' in some case studies to environmental and technical aspects, the usual objectives (e.g., Wagner and Raymond 2015). Furthermore, just a few parts brings the economics.

The Waste Management Policy Strategy

The term 'policy strategy' refers to the national planning stage development, with drivers, objectives and goals definitions, key-variables selection, standardized proxies and performance indicators assemble (existent), or its setup (new ones). The waste management plan shall include a complete information system to support the decisions (DSS) of the national regulatory agency, steering committees or the government. The European Commission is the only organism that issued aspecific manual on national WM planning for country members (Ljunggren 2000; Andersen et al. 2007; European Commission 2012).

The words like material flows and efficiency, environmentally sound waste, integrated system management, and holistic decision model reveal the intention of developed countries to reduce the relative waste production, and to the landfilling diversion. Meanwhile, a universal collection service and the adequate final disposal of waste is still a concern at DC Agenda (Carvalho et al. 2012; Bufoni et al. 2014; UNEP 2015).

This asymmetry is more evident in official reports published (European Environment Agency 2013; Corselli-Nordblad 2014; OECD 2015), and at the manuals, guidelines and recommendations issued by the countries commissions, regulatory agencies and international organizations (OECD 2007; European Comission 2008; USEPA 2014a; European Commission 2015). As result, the statistics show that, although "the effective technologies required to 'solve' the waste issues are largely already available, and have been much written about", while DC waste is changing in composition and increasing in production per capita resembling the developed ones, those cities rarely can adequate manage it, impacting health, the environment and the economy (Hoornweg and Bhada-Tata 2012; Wilson and Velis 2015).

In fact, waste management system in DC is still driven by sanitation and public health concerns (Bleck and Wettberg 2012; Bufoni et al. 2014), the first of five phases, historically associated with 1970s developed country waste management system (Marshall and Farahbakhsh 2013). Marshall and Farahbakhsh (2013) organize five historical drives associated with developed countries, from 1960s to nowadays: (1) Public Health; (2) Environment; (3) Resource Scarcity; (4) Climate Change; (5) Public concern and awareness. In their conclusions, however, the authors indicate that the current DC conditions are starkly different, because of the

> Rapid urbanization, soaring inequality, and the struggle for economic growth; varying economic, cultural, socioeconomic, and political landscapes; governance, institutional, and responsibility issues; and international influences have created locally specific, technical and non-technical challenges of immense complexity—(Marshall and Farahbakhsh 2013, p. 992)

Though, others studies suggest this list is far larger. The Guerrero et al. (2013) review on WM of 37 cities in 22 countries of 4 continents, per example, presents a 29 factors that influence the elements of WM systems and concludes for 5 critical aspects that would be addressed: (1) the role of each system stakeholder; (2) decisions complexity not only based on technologies; (3) financial support; (4) reliable data from a proper information channels; (5) university and research centers development.

Godfrey et al. (2013) concentrates efforts on political obstacles associated with other interests and the lack of knowledge of the political authorities to deliver and to support an efficient management system. The problem is that political environment determines the policy making and, in consequence, the regulatory framework. Bufoni et al. (2015) from a 2004–2014 sample of 432 large waste management projects deduce that a poor regulatory framework impact negatively the attractiveness of waste and waste-to-energy sectors, also commending a financial database building and standardization.

The DC regulatory shortcomings are well documented, and its implications are manifold. Examining South Korea's renewable energy deployment, Yoon and Sim (2015) list some causes for a policy failure that may hinder the success in four evaluation areas: (1) policy environment; (2) policy design; (3) policy implementation; (4) policy monitoring.

1. **Policy environment**
   1.1. Lopsided [others] political and industrial interests
   1.2. Lack of willingness to provide institutional support for deployment
   1.3. Physically and technologically good, but financially bad potentials for developments.
2. **Policy design**
   2.1. Problematic conceptualization and classification
   2.2. Disordered legal system and unsystematic national plans deployment
   2.3. Insufficient consideration before making the policy shift from feed-in-tariff to portfolio standards.
3. **Policy implementation**
   3.1. Insufficient financial support, and weak public relations, and marketing
   3.2. Inadequate coordination and lack of cooperation between government agencies.
4. **Policy monitoring, assessment, and feedback**
   4.1. Lack of systematic monitoring, regular evaluation, and performance auditing
   4.2. Lack of adequate feedback arrangements.

It is clear that these problems are not a renewable energy sector (including waste-to-energy) or a South Korea's privilege. Furthermore, besides not considering if this regulatory environment was well-designed, it does not have any conceptual problem, or lack of an adequate information system to assess its performance, the entire process in DC is deemed extremely dawdling (Massarutto 2007; Simões and Marques 2012). When the law or another instrument is deployed, if no other problem occurs, it can be already obsolete, and the process restarts. In some cases, a national policy on waste management can take decades. This issue was, per example, the case of Brazil (Brasil 2010).

A recent global report by United Nations Environment Programme (UNEP 2015) on waste management shows the importance of a coherent definition of the system aim according to the 'particular drivers'. In fact, the report and the WM literature considers that overall aims of a WM system change in time, depending on these drivers, but they are also country development dependent. Therefore, the UNEP report attests that

> It [the report] aims to provide assistance in identifying and implementing the appropriate policies and actions for the next steps in developing their own specific waste and resource management systems.
>
> It is, however, essential to highlight from the beginning that every situation is different. Thus there are no inherently 'right' or 'wrong' solutions, nor is it possible to provide a simple 'user manual' that will solve every problem (UNEP 2015).

This Global Waste Management Outlook (GWMO) report declares the fact that a sustainable WM is a challenge in developing countries (DC), mainly because of regulatory and financial constraints. The report recommends as analytical and conceptual frameworks: environmentally sound (ESWM) (COP 10 2011), the integrated sustainable (ISWM) (Guerrero et al. 2013), Life-Cycle Assessment (LCA) (Karmperis et al. 2013), and WM Hierarchy. However and again, according to the document, the lack of standards, measurement methodologies, and reporting systems still preclude the performance indicators and monitoring activities at those countries.

This chapter infers from this state of affairs that the waste management system in developed and developing countries are comparatively different, consequently, should also be the decision models. However, the review of the waste management literature indicates that both are prescriptive, managerial, based on performance and the classic model plan-do-check-review-revise (Gentil et al. 2010; Eriksson and Bisaillon 2011; Karmperis et al. 2013; Rashidi 2014; Laurent et al. 2014). Furthermore, a more recent study on the declared barriers of waste management projects implementation, suggests that a 'restrictions model' would be relevant and helpful to accelerate efficiency and efficacy of the WM system (Bufoni et al. 2016). The aid tool would permit the systematic and persistent search for systems performance constraints, in a considerably shorter time.

The summary of this part is that the waste management national plan is a complex system that depends on a multivariable set of conditions to work properly, from the environment to monitoring tools. These conditions are more easily found in developed countries. Their decision models are based on performance, life-cycle assessment and sophisticated information systems. The WM literature and the practitioners declare that such conditions in DC are rarely found. Hence, regulatory, financial and technological (transfer or development) barriers need attention.

In the interim, the developing countries will continue to choose the minimum capital expenses, the simplest and the more reliable and experimented feasible technology available to comply with a usually outdated, recent, precarious, and weak regulatory environment. In practical terms, this means public health drives, adequate final destination, low reduce-reuse-recycling rates and society awareness.

The choice trends for the coming years seem obvious and inevitable: landfill projects. This project type predominates, and on many occasions is the only mandatory 'treatment', equivocally declared adequate and sometimes even the 'recommended' one (Barton et al. 2008; Rogger et al. 2011). But not without consequences.

The Landfill Sites

Although the predominance of landfill projects worldwide, information about the waste management in DC, as we already told, is not easy to find. The projects design is usually associated with direct administration or bidding for a public concession to operate the landfill sites. Thus the information broadcast is transitory within the treaty process and/or simply not disclosure due to the government opacity.

Usually, the international reports about landfilling are statistical. It means data suffer some collection, analysis, interpretation or explanation to summarize the picture in a 'landscape,' 'outlook' or 'at a glance.' In the process much valuable information is lost or, depending on the report objective, diminished (Inanc et al. 2004). Even on OECD, US Protection or European Environment Agencies, it is not common to find details (OECD 2007; European Environment Agency 2013; USEPA 2014b).

One of the most consulted nonacademic sources for developing country waste management publications is the World Bank. The World Bank produces global (World Bank 2008; Hoornweg and Bhada-Tata 2012), blocks and continentals (Hoornweg and Giannelli 2007; World Bank 2011; Roos et al. 2012; Echart et al. 2012) and country (Hanrahan et al. 2006; World Bank 2010a, b) references. It includes not only many different approaches, from consumer patterns to technology costs but also several waste management industries like incineration, However, projects details are rare.

Therefore, the UNFCCC source of projects design documents, as part of Clean Development Mechanism (CDM) registration, increase in importance (UNFCCC 2015). With more than one thousand waste management projects, which 70 % are landfill projects, this mandatory process is reputed as one of the richest and most important information sources for research by waste management designers and practitioners in developing countries (Plochl et al. 2008; Bufoni et al. 2015). Therefore, the following conclusions are supported by 433 large projects analyses.

The findings show that the regulatory baseline of most countries is venting the landfill gas (LFG), although, the same regulatory theoretically recommended practice recognizes that send to a final destination only the residues not 'economic viable, or technological feasible' is the ideal. From more than 300 hundred landfill projects, only three of them have in site recycling or biodigesters facilities, possibly to the economic technical and regulatory barriers (Tayyeba et al. 2011). The rest usually declare itself planned to, in the first phase, collect the LFG and flare it, with some leachate treatment (Yue et al. 2007).

The second project phase selected would be purifying and reduce moisture of the biogas (half methane and half CO_2) to fuel engines to produce energy thru generators, or the upgrade purifying to a >99 % CH_4 gas content, usually to fuel a NGV engine, including for waste truck fueling purposes (DIESEL-NGV Otto cycle engine), or simply send it to the grid. In this phase, the project is exposed to many LFG production uncertainties and pitfalls (Han et al. 2009; El-Fadel et al. 2012).

However, the most common solution to electricity production is the use of biogas reciprocating engines. The mode is to use from 500 up to 1,500 kW engines brands as Stamford, MAN, MTU, Guascor, Marelli, ShangdongShengdong, Luoyang Zhongzhong, Caterpillar, and GE Jenbacher. These engines were designed specifically for biogas applications (50 % CH_4) and are characterized by a 'particularly high efficiency, low emissions, high durability and high reliability'.

Nevertheless, the engines are expensive and require some gas standards to work properly, which demands more costly equipment what, conversely, may be prohibitive for some projects. Therefore, many projects postpone sine die the inclusion of engines (phase 2). The business as usual (BAU) scenario is to use multiple smaller engines (3–32) due to increased flexibility in meeting a project's gas lifetime production curve (Bufoni et al. 2015).

However, let's draw some attention to LFG gas production uncertainties problem. The uncertainties, in this case, are threefold: (1) the inadequacy of gas emissions/production theoretical models; (2) efficiency of collection systems; (3) the unexpected and unpredictable site condition. All are interlinked and together created a relevant social opposition based on a symbolic, cultural, and ideological factors, and environmental awareness (NIMB—not in my backyard). The public perception is presumed to be a key element of the willing to accept new technologies (Massarutto 2007; Kardooni et al. 2016). Some investors complained about a 'counter-culture' on those projects and to suffer some technology prejudice.

Furthermore, the literature states that the adequate landfilling is always a better destination solution than controlled and open dumps, but still unsustainable. The problem is the incompatibility of its components lifetime (20–30 years) and the leachate and heavy metals lifecycles (110–250 years) (DoBerl et al. 2002; Malkow 2004). The difficulty is greater in countries in which the sites are numerous and older (developed ones), a situation that does not need to be replicated by DC, but it is the trend.

Supposing the adequate destination at developing countries are still a problem, it has many uncontrolled dumps, which the underground condition is a mystery, especially to the insufficient site prospection to determine the organic fraction and the moisture of the waste (Chung and Lo 2008; Siddiqui et al. 2013). Even the controlled ones conditions are difficult to predict since the machinery maintenance is humble and the aftercare efficiency is suspicion (Al-Yaqout and Hamoda 2002; El-Fadel et al. 2012; Bufoni et al. 2014).

As a result, the existent predictive models like UK GasSim, US landGEM, TNO, Mexico or UN FOD fail to hit the observed landfill gas production (Donovan et al. 2011; Aydi et al. 2015). Models are simplified reality representations, a reductionism of phenomena by philosophical systems. In other words, the models' technical parameters, called coefficients, are unable to capture the entire specificity of the site conditions.

Per example, due to the constant criticism about the too conservative predictive value of LandGEM model used by United States Environment Protection Agency (Wang et al. 2013), and the need for a better environment control, the agency established a pilot program from 2006 to 2026 building 51 bioreactors landfills in substitution to the traditional 'dry tomb'. A bioreactors landfill is defined as "any landfill cell where liquid, or air is injected in a controlled fashion into the waste mass to accelerate or enhance bio stabilization of the waste" (USEPA 2015). Even a more used model as the first order decay (FOD) of the United Nations has some statistically significant differences and can result in substantial planning errors about the production potential by the regulatory agencies (Amini et al. 2013; Loureiro et al. 2013).

The LFG collection system efficiency is also another issue. It is not because of soil cover fugitive emissions. Studies on the recovery efficiency of gas collection and retrieval system reported at most 85 %, but its effectiveness can fall up to 41 %. The default expected average is 70 % (Mønster et al. 2015; Aydi et al. 2015). These numbers are far from the alternative organic fraction treatment and gas recovery efficiency like anaerobic digesters (>99 %), maybe the reason why these is one of the commended landfilling diversion options.

In summary, the unplanned and unmanaged landfills predominate at developing countries. Thus, the site conditions, from the soil cover till the bottom liners are uncertain and also unpredictable. The operational risk associated with unexpected, and experienced underperformance, results in a partial or total projects withdrawal. Low compliance and enforcement baselines, skilled labor, and inadequate equipment maintenance complete the description of the current and possibly the future situations because the impacts will be felt in the post-closure monitoring period. A part of these upcoming issues are technological and depend on solutions research and development already in course, but yet not available. Nevertheless, another part is structurally sociopolitical and regulatory dependent.

The Increasing Costs

However, it is possible to notice that another part of the reported problems could be solved with a proper financial configuration and investment budget. In this section, we propose to discuss this issue. First, we should divide it into three approaches: (1) internal (financial), (2) external (economics) and (3) comparative risk-return analyses (benchmarking).

According to the World Bank, the waste management costs in most DC cities are the main yearly budget service expense of the municipalities. Also, the 2012 report points out the inequity of these distribution costs, where the operational expenses are concentrated at waste collection and transfer, usually to an inadequate final disposal. However, the bank also finds a tendency to increase and relocation of these resources to an investment in intermediate stations, more sophisticated waste treatment facilities and landfills improvement (Hoornweg and Bhada-Tata 2012).

The UNEP Global Waste Management Outlook highlights not only the financing importance and challenge but also the cost of inaction. The report concludes that the benefits greatly exceed the costs. However, out of the 500 largest cities of DC, only 20 % are deemed creditworthy because of poor banking service coverage or the high level of cities indebtedness (UNEP 2015). The document recommends as good practices "using performance-based contracting, capping fees, building up funds for maintenance and replacement and tightening policy obligations", but it emphasizes the intimate relationship with country income and the need for a huge investment finance in the low-income DC waste sector.

The UNFCCC financial barrier analyses of the CDM registration process models the sources regarding two components: the investment finance (CAPEX) and revenue/expenses (OPEX). The studies reveal that lasting, adequate and stable source of income is the most important factor that impacts the waste management projects feasibility (Bufoni et al. 2015; UNEP 2015; Polzin et al. 2015), and it is increasingly more challenging if activities are policy-driven (such as environmentally sound treatment and disposal), then where the service is needed (i.e., primary waste collection). Here grows the importance of the environmental economics knowledge.

The external approach is more regulation-oriented mainly because of the new preference for economic instruments design and deploys, although the strong criticism. The method fundamentals are the social profit, polluter pays principle, market failure, and others environmental economics concepts (Oliveira and Rosa 2003; Carvalho et al. 2012; Dubois 2013; Maier and Oliveira 2014; MacKenzie and Ohndorf 2016). The main instruments—standards, Pigouvian taxes, subsidies and tradable permits, are deemed complex and hard to fine tune (Baumol and Oats 1988; Pearce and Turner 1989). Thus, only recently the developing world is increasing these solutions design and use.

However, there is a consensus on the significant influence that the regulatory environment has on how financial conditions are set and the level of investment to support a WM system because the tariffs are usually determined by the local government and unable to easy fluctuate. Indeed, the majority of the cities has an "all inclusive" waste management service within urban property value tax, approach that do not foster the production reduction and the system management efficiency, like the modern charge models like the 'pay as you throw' (PAYT) (Ogawa 2000; Puig-Ventosa 2008; Manaf et al. 2009).

The landfills can be administered directly or granted, and there is no evidence support for a better arrangement recommendation for every situation. However, the revenue concession is usually in monthly and fixed value basis. The tentative to change to a more performance-based system resulted in public resistance, low adherence rates, and inadequate fees (Lohri et al. 2014). Furthermore, the citizen rarely agrees that it is a fair system, because of the procedure design failures, or substantives, as the perception of treatment inequity (Batllevell and Hanf 2008).

The last less obvious approach is the benchmarking. Straight to the point, according to the tests, the landfill projects in average has the second worst rate of return of the entire waste industry, with the higher standard deviation. Furthermore, the operational expenses do not differ per example from an incinerator or manure solution, though correlation to energy production is far harder to predict (R^2). All these conditions suggest the landfill projects are less financially attractive and economic efficient, although essential. Nevertheless, compared to other renewable energy sources, the waste management is in a clear competitive disadvantage, because of the dual management characteristics (energy production and manage the waste) (Bufoni et al. 2015).

In summary, the budget provision for management of waste is increasing, but still unsatisfactory. Indeed, the costs are proportionally increasing with the efficiency improvement. In this matter, landfill industry needs especial attention. Regulatory instruments that could reduce the systemic risk and guarantee a more favor financial condition and stability are wanted. However, the solution would be suited for waste industry, not for a generic renewable energy one.

The Change of Potential Emissions

The previous parts of this chapter led us to draw some conclusions about the developing countries landfilling changes, regarding how improvements of landfill machinery, chosen strategy models, gas recovery, and projects operation and maintenance shall impact the natural environment. For this purpose, we adopt the 2006 IPCC Guidelines for National Greenhouse Gas Inventories method for estimating emissions of methane from SWDS—solid waste disposal sites (IPCC 2006).

The method is used worldwide to estimate GHG emissions and it is the unique and official method in most developing countries (Amini et al. 2013). It uses the First Order Decay (FOD) model (Eq. 1), where CH_4 is generated by an exponential factor that describes the fraction of degradable material which each year is degraded into CH_4 and CO_2.

Equation 1—Decomposable Degradable Organic Carbon (DDOCm)

$$\mathrm{DDOCm} = W \cdot \mathrm{DOC} \cdot \mathrm{DOC_f} \cdot \mathrm{MCF}$$

where

DDOCm	mass of decomposable DOC deposited, Gg
W	mass of waste deposited, Gg
DOC	degradable organic carbon in the year of deposition, fraction, Gg C/Gg waste
$\mathrm{DOC_f}$	fraction of DOC that can decompose (fraction)
MCF	CH_4 correction factor for aerobic decomposition in the year of deposition (fraction).

It means that the methane generation is a function of the waste deposition (volume) and composition (type) because of the amount of decomposable degradable organic carbon (DDOCm) contained. It also depends on the waste control, placement and management practices at the disposal sites. The part of the waste that will decompose under aerobic conditions (before the conditions becoming anaerobic) in the SWDS is interpreted with the methane correction factor (MCF). The IPCC SWDS classification and correction factor according to site conditions is

1. **Anaerobic-managed solid waste disposal sites (MCF = 1.0)** These must have controlled placement of waste (i.e., waste directed to specific deposition areas, a degree of control of scavenging and a degree of control of fires) and will include at least one of the following: (i) cover material; (ii) mechanical compacting; or (iii) leveling of the waste.

2. **Semi-aerobic managed solid waste disposal sites (MCF = 0.5)** These must have controlled placement of waste and will include all of the following structures for introducing air to waste layer: (i) permeable cover material; (ii) leachate drainage system; (iii) regulating pondage; and (iv) gas ventilation system.
3. **Unmanaged solid waste disposal sites (MCF = 0.8)** Deep and/or with high water table: All SWDS not meeting the criteria of managed SWDS and which have depths of greater than or equal to 5 m and/or high water table at near ground level. Latter situation corresponds to filling inland water, such as a pond, river or wetland, by waste.
4. **Unmanaged shallow solid waste disposal sites (MCF = 0.4)** All SWDS not meeting the criteria of managed SWDS and which have depths of less than five meters.
5. **Uncategorized solid waste disposal sites (MCF = 0.6)** Only if countries cannot categorize their SWDS into above four categories of managed and unmanaged SWDS, the MCF for this category can be used.

Thus, the change consequences are evident. Since the SWDS are changing into anaerobic-managed solid disposal sites (MCF = 1.0), independent which type it are changing from, the methane production potential will invariably increase.

Some studies have already pointed to the potential increased emissions, trying to forecast the impacts depending on the policy and sites conditions. Loureiro et al. (2013) analyzed the possible results of climate change policies, plans, and the government programs in MSW emissions, establishing three different scenarios to estimate the emissions from 2005 until 2030 of Rio de Janeiro State in Brazil. The emissions and reduction potential in future scenarios (Table 1) were

- **Scenario A**—with no intervention to minimize GHG emissions;
- **Scenario B**—that includes already planned actions and policies. That was, 100 % collection 80 % methane recovery and controlled burn in flairs; and
- **Scenario C**—is where all preliminary planned and the government goals were archived with emission reduction by 65 % compared to 2005, eradicate the use of open dumps by 2014 and their remediation by 2016.

The study concludes that "there are many different procedures and technologies to treat solid waste, but in Brazil, landfilling seems to be a good solution for waste

Table 1 Emissions in future scenarios for solid waste in the state of Rio de Janeiro (GgCO2eq)

Emissions	Scenarios		
Year	A	B	C
2010	5152	4032	3771
2015	5890	3797	2248
2020	6620	3432	1501
2025	7415	2729	1695
2030	8314	1774	1874

Source Loureiro et al. (2013)

treatment in the coming years, to the extent that more economical and environmentally adequate solutions are not possible to implement." Note that the emissions reductions depend on 100 % collection and 80 % of the methane recovery or flare.

However, Bufoni et al. (2014) do not agree, arguing that economic and technological viable to be concepts that depend on the correct mix of incentives made available by an adequate and coherent waste management policy. Furthermore, as earlier already seen, the sound management, 100 % collection, and 80 % recovery scenario are very unlikely to occur. A simple visit to a sample of 7 alleged "landfills" at Rio de Janeiro State, and Marques (2013) did not find a single flare in order, broken and old equipment, poor management, and other operational problems.

Conclusions

At the first part of the chapter we briefly show the subject relevance and the main differences between the developed and developing countries waste destination and landfilling diversion, and how the waste literature itself recognizes, that system research would be more comprehensive, and it also would address the root of the problems.

The second part discusses the incongruence, incompatibility and, perhaps, the insufficiency of the simple reckoned global accepted performance models to, alone; solve the developing countries path obstacles to an "environmental sound waste management." The (many) problems constantly threats and many times prevent the plan's feasibility, which decisions process are sometimes biased and unbalanced. A side effect is the entire political and regulatory dynamics slowdown. We conclude for the trend of a new approach concentrating efforts in the systematical alleviation of constraints. The movement seems to have already been initiated, but not yet modeled or theoretically supported by United Nations.

Meanwhile, the third part presents an unpleasant figure of the current situation of the chosen strategy, planning and operation of national waste management systems. The public health care objective, associated with the financial and investment resources scarcity led to the apparently cheaper and backward alternatives. However, the flaw disappears if we consider the social cost of no action, post-closure expenses and in some specific cases the land value (hedonic). The technological uncertainties related to open dumps and unmanaged SWDS conditions and components lifespan are tough.

The financial consequences to the waste management sector are reflexive. The fourth part translates this uncertainties impact in currency effects. First regarding low attractiveness comparatively to others sectors rate of return and with the same risk. Second studies show while the operational expenses do not differ from other conversion technologies, there is a reasonable doubt about adequate, long-standing, and performance-based revenue accentuating the regulation inadequacy to remunerate the industry appropriately. The good news is that developing countries begin to develop suited waste management economic instruments and regulation. The same can be observed in quality and other parameters of waste-to-energy products like biodiesel, natural gas, and biocoal.

The landfill gas production potential increases with the most anaerobic-managed characteristics of the new modern disposal sites. However, the costs are strong and directly related to the location emissions. The fifth part conclusion is that to realize the expected emissions reduction the landfill operation request better attention, or the current practices will backfire.

References

Allesch A, Brunner PH (2014) Assessment methods for solid waste management: a literature review. Waste Manag Res 32:461–473. doi:10.1177/0734242X14535653

Al-Yaqout AF, Hamoda MF (2002) Report: management problems of solid waste landfills in Kuwait. Waste Manag Res 20:328–331

Amini HR, Reinhart DR, Niskanen A (2013) Comparison of first-order-decay modeled and actual field measured municipal solid waste landfill methane data. Waste Manag 33:2720–2728. doi:10.1016/j.wasman.2013.07.025

Andersen FM, Larsen H, Skovgaard M et al (2007) A European model for waste and material flows. Resour Conserv Recycl 49:421–435. doi:10.1016/j.resconrec.2006.05.011

Aydi A, Abichou T, Zairi M, Sdiri A (2015) Assessment of electrical generation potential and viability of gas collection from fugitive emissions in a Tunisian landfill. Energy Strateg Rev 8:8–14. doi:10.1016/j.esr.2015.06.002

Bartl A (2011) Barriers towards achieving a zero waste society. Waste Manag 31:2369–2370. doi:10.1016/j.wasman.2011.09.013

Bartl A (2014) Moving from recycling to waste prevention: a review of barriers and enables. Waste Manag Res 32:3–18. doi:10.1177/0734242X14541986

Barton JR, Issaias I, Stentiford EI (2008) Carbon–making the right choice for waste management in developing countries. Waste Manag 28:690–698. doi:10.1016/j.wasman.2007.09.033

Batllevell M, Hanf K (2008) The fairness of PAYT systems: some guidelines for decision-makers. Waste Manag 28:2793–2800. doi:10.1016/j.wasman.2008.02.031

Baumol W, Oats W (1988) The theory of environmental policy, 2nd edn. Cambridge University Press

Bleck D, Wettberg W (2012) Waste collection in developing countries–tackling occupational safety and health hazards at their source. Waste Manag 32:2009–2017. doi:10.1016/j.wasman.2012.03.025

Brasil (2010) LEI N° 12.305, DE 2 DE AGOSTO DE 2010. Institui a Política Nacional de Resíduos Sólidos. Congresso Nacional

Brunner PH, Fellner J (2007) Setting priorities for waste management strategies in developing countries. Waste Manag Res 25:234–240. doi:10.1177/0734242X07078296

Bufoni AL, Carvalho MS, Oliveira LB, Rosa LP (2014) The emerging issue of solid waste disposal sites emissions in developing countries: the case of Brazil. J Environ Prot (Irvine, Calif) 5:886–894. doi:10.4236/jep.2014.510090

Bufoni AL, Oliveira LB, Rosa LP (2015) The financial attractiveness assessment of large waste management projects registered as clean development mechanism. Waste Manag 43:497–508. doi:10.1016/j.wasman.2015.06.030

Bufoni AL, Oliveira LB, Rosa LP (2016) The declared barriers of the large developing countries waste management projects: the STAR model. Waste Manage 52. http://doi.org/10.1016/j.wasman.2016.03.023

Carvalho MS, Rosa LP, Bufoni AL, Ferreira AC de S (2011) The issue of sustainability and disclosure. a case study of selective garbage collection by the urban cleaning service of the city of Rio de Janeiro, Brazil–COMLURB. Resour Conserv Recycl 1030–1038. doi:10.1016/j.resconrec.2011.05.015

Carvalho MS, Rosa LP, Bufoni AL, Oliveira LB (2012) Putting solid household waste to sustainable use: a case study in the city of Rio de Janeiro, Brazil. Waste Manag Res

Chung SS, Lo CWH (2008) Local waste management constraints and waste administrators in China. Waste Manag 28:272–281. doi:10.1016/j.wasman.2006.11.013

Ciplak N, Barton JR (2012) A system dynamics approach for healthcare waste management: a case study in Istanbul Metropolitan City, Turkey. Waste Manag Res 30:576–586. doi:10.1177/0734242X12443405

COP 10 (2011) BC-10/2: strategic framework for the implementation of the Basel Convention for 2012–2021

Corselli-Nordblad L (2014) In 2012, 42 % of treated municipal waste was recycled or composted. Eurostat News Release 48–50

DoBerl G, Huber R, Brunner PH et al (2002) Long-term assessment of waste management options—a new, integrated and goal-oriented approach. Waste Manag Res 20:311–327. doi:10.1177/0734247X0202000402

Donovan SM, Pan Jilang, Bateson T et al (2011) Gas emissions from biodegradable waste in United Kingdom landfills. Waste Manag Res 29:69–76. doi:10.1177/0734242X10389510

Dubois M (2013) Towards a coherent European approach for taxation of combustible waste. Waste Manag 33:1776–1783. doi:10.1016/j.wasman.2013.03.015

Echart J, Ghebremichael K, Khatri K, et al (2012) The future of water in African Cities : why waste water? Integrated Urban Water Management, Background Report

El-Fadel M, Abi-Esber L, Salhab S (2012) Emission assessment at the Burj Hammoud inactive municipal landfill: viability of landfill gas recovery under the clean development mechanism. Waste Manag 32:2106–2114. doi:10.1016/j.wasman.2011.12.027

EPA (2011) Criteria for municipal solid waste landfills. Code of Federal Regulations, Environmental Protection Agency, Protection of Environment

Eriksson O, Bisaillon M (2011) Multiple system modelling of waste management. Waste Manag 31:2620–2630. doi:10.1016/j.wasman.2011.07.007

European Commission (2008) Directive 2008/98/EC on waste and repealing certain directives. European Parliament and of the Council

European Commission (2012) Preparing a waste management plan

European Commission (2015) Waste prevention programmes. http://scp.eionet.europa.eu/facts/WPP. Accessed 17 Dec 2015

European Environment Agency (2013) Managing municipal solid waste—a review of achievements in 32 European counties

Gentil EC, Damgaard A, Hauschild M et al (2010) Models for waste life cycle assessment: review of technical assumptions. Waste Manag 30:2636–2648. doi:10.1016/j.wasman.2010.06.004

Godfrey L, Scott D, Trois C (2013) Caught between the global economy and local bureaucracy: the barriers to good waste management practice in South Africa. Waste Manag Res 31:295–305. doi:10.1177/0734242X12470204

Guerrero LA, Maas G, Hogland W (2013) Solid waste management challenges for cities in developing countries. Waste Manag 33:220–232. doi:10.1016/j.wasman.2012.09.008

Han H, Qian G, Long J, Li S (2009) Comparison of two different ways of landfill gas utilization through greenhouse gas emission reductions analysis and financial analysis. Waste Manag Res 27:922–927. doi:10.1177/0734242X09349762

Hanrahan D, Srivastava S, Ramakrishna AS (2006) Improving management of municipal solid waste in India : overview and challenges

Hoornweg D, Bhada-Tata P (2012) What a waste : a global review of solid waste management

Hoornweg D, Giannelli N (2007) Managing municipal solid waste in latin America and the Caribbean : integrating the private sector. Harnessing Incentives

Inanc B, Idris A, Terazono A, Sakai S (2004) Development of a database of landfills and dump sites in Asian countries. J Mater Cycles Waste Manag. doi:10.1007/s10163-004-0116-z

IPCC (2006) 2006 IPCC guidelines for national greenhouse gas inventories, vol 5. Waste

Kardooni R, Yusoff SB, Kari FB (2016) Renewable energy technology acceptance in Peninsular Malaysia. Energy Policy 88:1–10. doi:10.1016/j.enpol.2015.10.005

Karmperis AC, Aravossis K, Tatsiopoulos IP, Sotirchos A (2013) Decision support models for solid waste management: review and game-theoretic approaches. Waste Manag 33:1290–1301. doi:10.1016/j.wasman.2013.01.017

Laurent A, Bakas I, Clavreul J et al (2014) Review of LCA studies of solid waste management systems–part I: lessons learned and perspectives. Waste Manag 34:573–588. doi:10.1016/j.wasman.2013.10.045

Ljunggren M (2000) Modelling national solid waste management. Waste Manag Res 18:525–537. doi:10.1177/0734242X0001800603

Lohri CR, Camenzind EJ, Zurbrügg C (2014) Financial sustainability in municipal solid waste management—costs and revenues in Bahir Dar, Ethiopia. Waste Manag 34:542–552. doi:10.1016/j.wasman.2013.10.014

Loureiro SM, Rovere ELL, Mahler CF (2013) Analysis of potential for reducing emissions of greenhouse gases in municipal solid waste in Brazil, in the state and city of Rio de Janeiro. Waste Manag 33:1302–1312. doi:10.1016/j.wasman.2013.01.024

MacKenzie IA, Ohndorf M (2016) Coasean bargaining in the presence of Pigouvian taxation. J Environ Econ Manage 75:1–11. doi:10.1016/j.jeem.2015.09.003

Maier S, Oliveira LB (2014) Economic feasibility of energy recovery from solid waste in the light of Brazil's waste policy: the case of Rio de Janeiro. Renew Sustain Energy Rev 35:484–498. doi:10.1016/j.rser.2014.04.025

Malkow T (2004) Novel and innovative pyrolysis and gasification technologies for energy efficient and environmentally sound MSW disposal. Waste Manag 24:53–79. doi:10.1016/S0956-053X(03)00038-2

Manaf LA, Samah MAA, Zukki NIM (2009) Municipal solid waste management in Malaysia: practices and challenges. Waste Manag 29:2902–2906. doi:10.1016/j.wasman.2008.07.015

Marques F (2013) The operational proceeding evaluations of sanitary landfills of Rio de Janeiro State. Universidade do Estado do Rio de Janeiro

Marshall RE, Farahbakhsh K (2013) Systems approaches to integrated solid waste management in developing countries. Waste Manag 33:988–1003. doi:10.1016/j.wasman.2012.12.023

Massarutto A (2007) Municipal waste management as a local utility: options for competition in an environmentally-regulated industry. Util Policy 15:9–19. doi:10.1016/j.jup.2006.09.003

Mønster J, Samuelsson J, Kjeldsen P, Scheutz C (2015) Quantification of methane emissions from 15 Danish landfills using the mobile tracer dispersion method. Waste Manag 35:177–186. doi:10.1016/j.wasman.2014.09.006

OECD (2015) Municipal waste

OECD (2007) Guidance manual for the implementation of the OECD recommendation C(2004)100 on environmentally sound management (ESM) of waste

Ogawa H (2000) Sustainable solid waste management in developing countries. In: Organization WH (ed) 7th ISWA international congress and exhibition. Kuala Lumpur, Malaysia

Oliveira LB, Rosa LP (2003) Brazilian waste potential: energy, environmental, social and economic benefits. Energy Policy 31:1481–1491. doi:10.1016/S0301-4215(02)00204-5

Pearce DW, Turner RK (1989) Economics of natural resources and the environment. Johns Hopkins University

Plochl C, Wetzer W, Ragossnig A (2008) Clean development mechanism: an incentive for waste management projects? Waste Manag Res 26:104–110. doi:10.1177/0734242X07087947

Polzin F, Migendt M, Täube FA, von Flotow P (2015) Public policy influence on renewable energy investments—a panel data study across OECD countries. Energy Policy 80:98–111. doi:10.1016/j.enpol.2015.01.026

Puig-Ventosa I (2008) Charging systems and PAYT experiences for waste management in Spain. Waste Manag 28:2767–2771. doi:10.1016/j.wasman.2008.03.029

Rashidi R (2014) A review of performance management system. Int J Acad Res 7:210–214

Rogger C, Beaurain F, Schmidt TS (2011) Composting projects under the clean development mechanism: sustainable contribution to mitigate climate change. Waste Manag 31:138–146. doi:10.1016/j.wasman.2010.09.007

Roos K, Ru J, Zhou W (2012) An innovative and cost-effective solution for livestock waste management in China, Thailand and Vietnam

Siddiqui FZ, Zaidi S, Pandey S, Khan ME (2013) Review of past research and proposed action plan for landfill gas-to-energy applications in India. Waste Manag Res 31:3–22. doi:10.1177/0734242X12467066

Simões P, Marques RC (2012) Influence of regulation on the productivity of waste utilities. What can we learn with the Portuguese experience? Waste Manag 32:1266–1275. doi:10.1016/j.wasman.2012.02.004

Singh J, Laurenti R, Sinha R, Frostell B (2014) Progress and challenges to the global waste management system. Waste Manag Res 32:800–812. doi:10.1177/0734242X14537868

Tayyeba O, Olsson M, Brandt N (2011) The best MSW treatment option by considering greenhouse gas emissions reduction: a case study in Georgia. Waste Manag Res 29:823–833. doi:10.1177/0734242X10396119

UNEP (2015) Global waste management outlook (GWMO). United Nations Environment Programme

UNFCCC (2015) CDM project search. https://cdm.unfccc.int/Projects/projsearch.html. Accessed 18 Dec 2015

USEPA (2014a) Municipal solid waste generation, recycling, and disposal in the United States Tables and Figures for 2012. 63

USEPA (2014b) Municipal solid waste generation, recycling, and disposal in the United States: Facts and Figures for 2012. Washington

USEPA (2015) Bioreactors, municipal solid waste. http://www3.epa.gov/epawaste/nonhaz/municipal/landfill/bioreactors.htm. Accessed 25 Dec 2015

Wagner TP, Raymond T (2015) Landfill mining: case study of a successful metals recovery project. Waste Manag. doi:10.1016/j.wasman.2015.06.034

Wang X, Nagpure AS, Decarolis JF, Barlaz MA (2013) Using observed data to improve estimated methane collection from select U.S. landfills. Environ Sci Technol 47:3251–3257. doi:10.1021/es304565m

Wilson DC, Velis CA (2015) Waste management—still a global challenge in the 21st century: an evidence-based call for action. Waste Manag Res 33:1049–1051. doi:10.1177/0734242X15616055

World Bank (2008) Solid waste management holistic decision modeling

World Bank (2011) Solid waste management in Bulgaria, Croatia, Poland, and Romania : a cross-country analysis of sector challenges towards EU harmonization

World Bank (2010b) Brazil low carbon case study : waste

World Bank (2010a) Establishing integrated solid waste management in the large cities of Pakistan Multan : comprehensive scope evaluation report

Yoon J-H, Sim K (2015) Why is South Korea's renewable energy policy failing? a qualitative evaluation. Energy Policy 86:369–379. doi:10.1016/j.enpol.2015.07.020

Yue D, Xu Y, Mahar RB et al (2007) Laboratory-scale experiments applied to the design of a two-stage submerged combustion evaporation system. Waste Manag 27:704–710. doi:10.1016/j.wasman.2006.04.017

Microbial Fuel Cells in Solid Waste Valorization: Trends and Applications

R.A. Nastro, G. Falcucci, M. Minutillo and E. Jannelli

Abstract In recent years, biomass valorization (and, in general, waste treatment) and FC technology met in the so-called bioelectrochemical systems (BESs). BESs take advantage of biological capacities (microbes, enzymes, plants) for the catalysis of electrochemical reactions. They mainly include micro-electrolysis Cell (MECs) and microbial fuel cells (MFCs). While MECs can produce valuable compounds (like H_2, CH_4, etc.), providing a suitable potential at the electrodes, MFCs do not need any energetic input to convert chemical energy (stored in organic compounds) into electric power. In this "biologically-based-fuel–cells," the fuel is made by different sources of organic compounds. Landfill leachate, municipal and agro-industrial wastewaters, sediments, solid organic wastes can be source of electric power and commodity chemicals. The use of MFC technology to waste treatment and valorization is, maybe, the most promising application of this newborn technology. Even though many researchers proved the reliable utilization of liquid waste as fuel in scaled MFCs, few attempts to apply MFCs to solid waste valorization have been done. In this paper, recent studies about the application of MFCs to solid substrates treatment and valorization and the contribution that BESs and MFC in particular could give to the development of a more sustainable waste management.

Keywords Organic fraction of municipal solid waste · Microbial fuel cell · Waste treatment · Waste-to-energy technology

G. Falcucci
Department of Engineering of the Enterprise Mario Lucertini, University of Rome Tor Vergata, Rome, Italy

R.A. Nastro (✉) · M. Minutillo · E. Jannelli
Department of Engineering, Parthenope University of Naples, Naples, Italy
e-mail: r.nastro@uniparthenope.it

D. Sengupta and S. Agrahari (eds.), *Modelling Trends in Solid and Hazardous Waste Management*, DOI 10.1007/978-981-10-2410-8_9

Introduction

It has been estimated that waste management contributes for about 3–5 % to green house gases (GHGs) emission being responsible for GHGs emission, mainly due to CH_4, CO_2, and N_2O escapes from open dumps. Additional CO_2 emissions are from upstream processes like waste collection and transportation (UNEP 2010). Nevertheless, an adequate waste management can save or reduce GHGs emissions in different ways: by primary materials avoidance through material recovery from waste, by energy production, by carbon storing in landfills and through the application of compost to soils. If we consider the recommendations by internationally recognized institutions, the future waste management should be essentially focused on the 3R concept (Reduce, Reuse, and Recycle), cleaner productions, circular economy establishment, waste prevention and, finally, the transformation of waste into a source of energy and materials (UNEP 2010). Besides of ecosystems alteration, air, water, and soil pollution, an inadequate waste management can represent a real threaten to human health. If toxic waste can have undiscussed effects on human health, the influence of Municipal Solid Waste (MSW) dumping and incineration on the population living near waste treatment facilities has not been fully clarified. Some studies gave evidence of a correlation between the proximity of local population at OFMSW facilities and few types of cancers, congenital anomalies, and low weight at birth (Rushton 2003). Nevertheless, the effects seem to vary according to the studied population, so a different approach could be used in the epidemiologic surveys in order to clarify if even the MSW treatment can be correlated without any doubt with human diseases (Giusti 2009; Porta et al. 2009). The problem related to an inadequate waste treatment is of a particular importance in developing Countries, where limited resources are destined to waste management. For this reason, in that Countries, dumping represents the most commonly used disposal method (often with no proper control) and consequent air, soil, and water pollution. It is clear that waste management represents one of the main issues mankind has to face nowadays. Nevertheless, waste can be a resource.

Waste: A Resource

Waste is a resource not only because of materials recovery (glass, metals, fibers, and plastics) and energy, but also because of oil saving. If we consider just the organic waste from agriculture (crop residues), the global energy that could be produced is estimated to be about of 50 billion tons of oil equivalent (UNEP 2010). The major issue is: *how can we exploit this resource minimizing the environmental impacts and costs*? UNEP indicates as a necessary prerequisite for an effective energy generation an adequate separation between organic and non organic waste: the organic residues are, in fact, responsible for the compromission of thermal technology effectiveness in terms of energy produced, besides of the GHGs emissions.

Organic Waste Treatment and Valorization

If we consider just the Organic Fraction of Municipal Solid Waste (OFMSW), the most commonly used technologies for its treatment and valorization are the Anaerobic Digestion (AD) and composting (UNEP 2010). These two waste treatments differ essentially for the microbial metabolism they are based on. Anaerobic Digestion (AD) is based on anaerobic microorganisms metabolism, with particular regards to methanogenic bacteria which can produce CH_4 from CO_2 to H_2 (hydrogenotrophs) or from CH_3COOH (acetoclastics). AD requires an appropriate temperature to occur: generally a temperature of 35 °C or 50–55 °C is realized in the reactor, even if a psicotrophic process is also possible (10–20 °C). As a result of the anaerobic digestion, a biogas rich in CH_4 is produced while the resulting digestate is very often aerobically stabilized.

If AD needs energetic inputs (mainly to keep a constant temperature and leachate recirculation) a successful composting procedure needs oxygen insufflation sufficient to sustain the aerobic microorganisms and, at that same time, inhibiting anaerobic bacteria. If small amounts of residues can be easily composted, large-scale composting requires mechanical aeration, i.e., energetic imputs, varing according to the technology used (approximately 40–70 kW/t of waste) (Faaij et al. 1998). This energy is normally provided to the system, but facilities combining AD and a following digestate aerobic stabilization can provide the energy needed for the composting process from self-supplied methane. It has been estimated that if 25 % or more of the waste is anaerobically digested, whole treatment system can be self-sufficient (UNEP 2010). The main "product" of composting is the stabilized organic matter used, if free from contaminants, as a soil conditioner.

Compost in field application is assumed to reduce the utilization of synthetic fertilizer (about 20 % according to IPCC (2014). On the whole, in-field soil conditioner application has positive impact on GHGs emission from primary production (fertilazers production), but also on N_2O emissions from soil and reduced irrigation, pesticides and tillage (Favoino and Hoggs 2008; Faaji 2006). Small, low-technology facilities handling only yard waste are inexpensive and generally problem-free. For developing Countries, the low cost and simplicity of composting make small-scale composting a promising solution (Faaji 2006). BESs (and, among them, MFCs) could represent a potentially low cost and effective technology for waste and wastewater technology and an important tool for waste management in poor Countries.

Bioelectrochemical Systems (BESs)

The ability of microorganisms to use inorganic molecules as electronic acceptor is widely spread in the environment: *Ferribacterium limneticum* and *Geobacter* spp, for example, are able to use Fe^{3+} as electronic acceptor, producing Fe^{2+}.

This process is normally associated with energy production in microbial cells (Coates et al. 2001; Cummings et al. 1999). BESs represent the attempt to manipulate this natural process to obtain electric power directly from the microbial metabolism, without any combustion. As based on microbial metabolism, BESs can be virtually fed with any organic compound, wastewater, urine, food waste and, even recalcitrant compounds like heavy metals, pesticides, hydrocarbons, dyes (ElMekawy et al. 2015; Nastro 2014; Pant et al. 2010; Shaoan and Chen 2011; Morris et al. 2009) and used with different purposes. BESs include microbial fuel cells (MFCs) and microbial electrolysis cells (MECs). microbial fuel cells (MFCs) are generally thought to provide electric power for small devices in deep see beds or remote areas using sediments and even urine or insects as feedstock (Jungh et al. 2014; Ieropoulos et al. 2005; Logan and Regan 2006). Recently, the reliable utilization of MFCs to charge the battery of a mobile has been demonstrated by Ieropoulos et al. (2013). Unlike MFCs, MECs require the set of an external potential at the electrodes in order to drive the electrochemical reactions to the synthesis of commodity compounds like caustic soda, hydrogen, methane. In the course of time, many different systems have been developed and tested all over the world. In general, different configurations are possible according to the geometry of the reactor, the materials used at the electrodes and for the chamber setup, the presence/absence of a cationic exchanging membrane, the application of an external potential, the nature of cathodic reactions, etc. A large overview of BESs has been given by Rabaey and Rozendal (2010) (Fig. 1).

As biomass-based systems, BESs are considered carbon neutral (Oh et al. 2010) and the biotransformation of organic matter into chemicals by microbial metabolism allows CO_2 emissions avoidance from primary production. Moreover, considering that MFCs do not involve CH_4 production and combustion, the environmental advantages linked to a future development of this technology could compensate even a higher cost of production (Pant et al. 2011). In order to drive BESs technology to a full in-field application, the setup of electrodes materials and cell layouts are of primary importance to obtain adequate performances in terms of power density, organic load removal and chemicals production yield (Wei et al. 2011; Nastro 2014; Sleutels et al. 2012; Pant et al. 2012). As the setup of a new technology cannot exempt from the evaluation of the environmental impacts related to the whole life cycle and MFCs scientists are working also in that direction: recently, a stack of MFCs was designed and set up with all biodegradable products, opening the possibility to setup energy devices that could degrade harmlessly into the surroundings to leave no trace when their mission is complete (Winfield et al. 2015).

Application of BESs to Waste Management and Bioenergy

BESs have the potential to play a major role in developing sustainable waste recycling systems, with reduced use of energy and, at the same time, generating useful chemicals. For this reason, the number of liquid and solid waste from

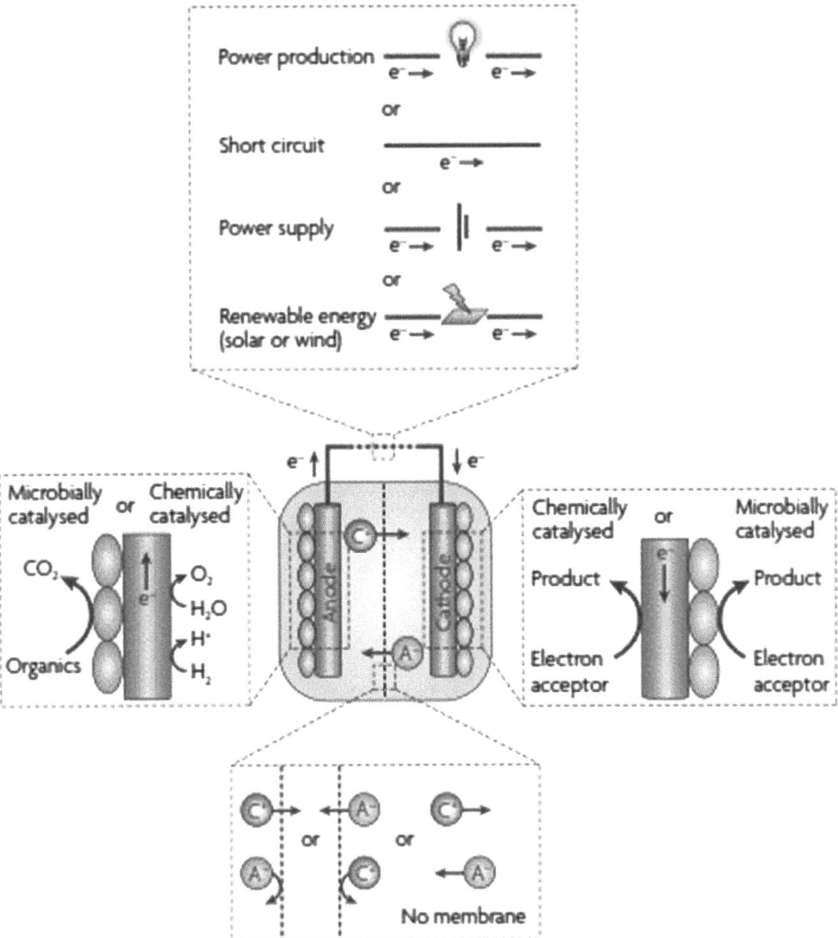

Fig. 1 A high-level overview of the concepts associated with bioelectrochemical systems (Rabaey and Rozendal 2010)

agro-industrial processes used as feedstock in BESs is increasing and increasing: a short list of agricultural substrates used in MFCs, with relative cells performances, is reported in Table 1. Among the other chemicals, hydrogen is maybe the best candidate as fuel for future technologies even because it can be the result of different processes catalyzed by bacteria like dark fermentation and bioelectrolysis (Chandrasekhar 2015). An intense research is actually being carried out to test biohydrogen production by micro-electrogenesis from different complex substrates like glycerol, milk, and starch in sight of a possible wide utilization of MECs to agro-industrial wastewaters, with biohydrogen or other chemicals production (Montpart 2015; Sleutels et al. 2012; Pant et al. 2012). Even if based on microbial metabolism, MFCs and MECs differ in many aspects, layouts and materials first of

Table 1 Some examples of food industry-based wastewater used as anolytes in MFCs and their respective performances (from Makawi et al. 2015 mod.)

Wastewater	MFC configuration	Concentration (in terms of COD mg/l)	Anode volume (ml)	Power density (PD)	CE (%)	ζCOD
Slaughter house wastewater	Dual chambered MFC	4850	125	578 mW/m^2	64 ± 2 %	93 ± 1 %
Animal carcass wastewater	Tubular air-cathode MFC	11,18	750	2.19 W/m^3	00.25.00	51.06.00
Swine wastewater	Dual chambered MFC	8320	250	45 mW/m^2	a	Not given
Swine wastewater	Single chambered cuboid MFC	8270	a	228 mW/m^2	a	84
Beer brewery wastewater	Single chambered membrane-free MFC	2239	a	483 mW/m^2	38	87
Beer brewery wastewater	Dual chambered MFC	a	200	16.29 mW/m^2	a	a
Brewery Wastewater	Single chambered MFC	1501	100	669 mW/m^2	10	20.7
Brewery wastewater	Tubular air-cathode MFC	2125	170	96 mW/m^2	28	93
Brewery wastewater	Three chamber reactor	2850	1200	a	a	54.2
Barley processing wastewater	Single chambered MFC	1200	100	a	1.65	95.11
Brewery diluted with domestic wastewater	Single chambered MFC	784	100	30 mW/m^2	a	90.4
Dairy wastewater	Single chambered MFC	3700	480	1.1 W/m^3 (~36 mW/m^2)	7.5	95.49
Dairy wastewater	Annular single chamber MFC	1000	90	20.2 W/m^3	26.87	91
Dairy wastewater	Dual chambered MFC	1600	300	161 mW/m^2	a	90

[a]Indicates the data not available from the cited reference

all. Cusick et al. (2010) carried out an interesting monetary evaluation of both MFC and MEC technologies applied to winery wastewater treatment concluding that energy recovery and organic removal from wastewater are more effective with MFCs than MECs, but hydrogen production from wastewater fed MECs can be cost effective. Besides of biohydrogen, the possibility to produce methane by MECs has being widely explored, with some encouraging results (Chandrasekhar et al. 2015; Villano et al. 2011). Even in this case, authors report a limitation of the performances linked to internal losses (high internal resistance, electrodes losses, etc.) that need to be overcome (Van Eerten-Jansen et al. 2012). MECs seems to be ready for a practical application to wastewater valorization (Sleutels et al. 2012) and, in fact, in recent years scaled prototypes for both municipal and industrial wastewaters valorization were set up in different countries allowing hydrogen peroxide, biohydrogen, methane, and caustic soda production using the chemical energy stored in municipal and industrial wastewaters. Nevertheless, the organic load removal from wastewater is still to be improved, being, for instance, up to 62 % in terms of COD removal in winery wastewaters, less than achieved with an activated sludge treatment plant (EU Commission 2013).

MFCs and Solid Waste Valorization

If a wide number of papers deal with the set up and the study of MFCs fed with wastewaters, few attempts to apply this newborn technology to solid organic waste treatment have been carried out. Mohan and Chandrasekhar published in 2011 a first paper dealing with the study of operational factors affecting the performances of MFCs fed with canteen food waste, focusing on electrodes distance and feedstock pH. Since then, other researchers started working on the application of MFCs to the Organic Fraction of Municipal Solid Waste (OFMSW) using different approaches: presence/absence of an inoculum, electrodes geometry, pH, temperature, oxygen availability, electrodes distance (Nastro et al. 2013; El-Chakhtoura 2014; Karluvali et al. 2015). Even if with different outcomes (Table 1), all researches confirm the effectiveness of MFC technology as tool for energy recovery and organic load removal from OFMSW, with particular regards to a low-temperature process (about 25 °C). Recently, Solid phase MFC (SMFCs) have been tested also in combination with a composting process of soybean, rice husk, leaf mold, and coffee residues used to prepare mixture with different C/N ratios (Wang et al. 2015). According to the authors, it is possible to combine the composting process with energy recovery by MFCs as the highest power density (Table 2) was achieved when the solid mixture C/N ratio was 30/1, able to sustain also a composting process. But SMFC are being tested also with agro-industrial substrates like Dried Distilled Grains with Solubles (DDGS), deriving from whisky production. Even in this case, laboratory scale tests confirmed the possibility to use MFC for waste valorization (Table 2). Further studies by Mohan gave evidence of the possible combination of MFC and biohydrogen production in a two-stage process (Chandrasekhar and Mohan 2014). Previously, Higgins et al. (2013) published an interesting research about the combination of MFCs and AD for solid organic waste treatment and valorization. If MFCs are fed with not pretreated OFMSW, they can represent a preliminary stage before a dark fermentation for

Table 2 Some examples of solid substrates in MFCs, with cells performances

Solid waste	MFC configuration	Anode volume (ml)	Power density	ΔCOD_{sol}	CE (%)	References
OFMSW	Single chamber	1000	Up to 13 mW/m^2 kg	Up to 80.4 %	Up to 5 %	Nastro et al. (2015a)
OFMSW	Single chamber	28	Up to 123 ± 41 mW/m^2	>86 %	Up to 24 ± 5 %	El-Chakhtoura et al. (2014)
OFMSW	Single chamber	250	Up to 47.6 mW/m^2	Up to 52.8 %	Up to 4.9 %	Karluvali et al.(2015)
Compost	Single chamber	200	Up to 4.6 mW/m^2	–	–	Wang et al.(2015)
DDGS	Single chamber	500	146 ± 11 mW/kg	–	–	Nastro et al. (2015b)
Composite food waste	Single chamber	500	8.8 mW/m^2kg	Up to76 %	n.r.	Chandrasekhar et al. (2011)

biohydrogen production or a methanogenesis. MFCs can also be used as treatment for the anaerobic digestion leachate, still rich in organic matter: in this case, MFC treatment is placed downstream. So, it is more likely that the future of MFCs and MECs in solid waste management will be essentially linked to the optimization of biohydrogen and biomethane production, besides of energy recovery from biomass and leachate (Premier et al. 2012). With the overcome of issues related to the scaling-up in terms of electrodes materials and geometry as well as operational parameters (temperature, retention time, pH, etc.) a full in-field application will be possible and, then, all the potentialities of MFCs will be completely and definitely explored.

Towards Scaling-up: Modeling MFCs Performances

Like other BESs, MFCs performances are heavily affected by several operational parameters such as reactor configurations and scales, electrode materials, electrode surface areas and the nature of electron donors, if present. If a wide range of data about specific parameters are actually available in lab-scale experiment, few studies are available about the dynamics of the wide range of biotic and abiotic parameters affecting MFCs power production in waste treatment. Nevertheless, the main bottlenecks actually limiting the performances of scaled MFCs seem to reside in the interactions among the above-cited factors. According to some authors, mathematical modeling can represent a powerful platform, helping researchers in investigating the synergistic effects of multiple parameters, (including biofilm composition and structure, redox mediator transferring, substrate utilization rate, etc.) on MFCs performances in waste management (Recio-Garrido et al. 2016). In general, mathematical modeling is performed in two approaches: engineering modeling, based on Differential Equations (DEs) implementing engineering/physical/biochemical/electrochemical laws governing the system processes, and a statistical modeling, based on the analysis of experimental measurements (Luo et al. 2016). In this context, the development and utilization of appropriate softwares are an important prerequisite. An example of MFCs modeling framework is reported in Fig. 2. In it, the authors implements electrochemical, biological, and structural MFCs parameters using both data from the literature (to elaborate an expected scenario) and from experimental activities (real scenario) by the use of Excel®, Mathlab®, and COMSOL® Multiphysics softwares (Oyetunde et al. 2013). Moreover, complex dynamic models are being developed in order to optimize and control MFCs and, in general, BESs processes (Recio-Garrido et al. 2015).

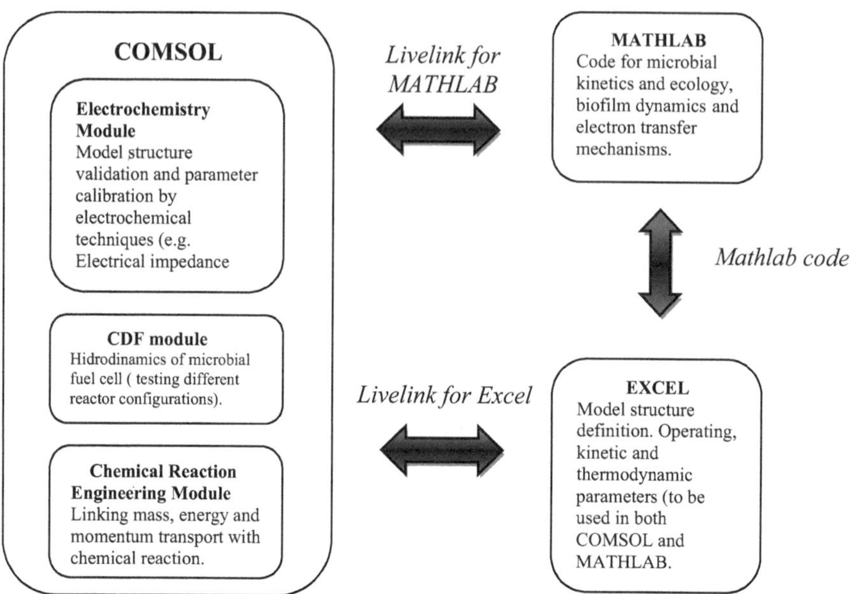

Fig. 2 Framework for investigating bioelectrochemical systems using COMSOL. Multiphysics (Oyetunde T et al. 2013)

Conclusions

The extraction of useful commodity chemicals, together with the production of energy, from any kind of organic waste and leftovers, is becoming more and more popular: it is a sustainable way to mitigate global warming, diversify energy sources and obtain chemicals, pharmaceuticals and food additives of high added value. BESs can represent a sustainable tool in waste management and bioenergy sector and recent researches confirm the high potentiality of such systems in terms of substrates to be treated and by-product that can be recovered, turning the "waste" into "resource." Among BESs, MFCs represent a possible alternative/integration to the AD and recent studies confirm the possibility to realize multistep waste treatment systems working at a temperature ranging from 20 to 25 °C, with energy saving and biohydrogen/methane/electric power recovery. A strong research is being carried out in different disciplines to take BESs to an in-field application and the utilization of these systems in waste valorization and treatment is not so far.

References

Chandrasekhar K, Mohan SV (2014) Induced catabolic bio-electrohydrolysis of complex food waste by regulating external resistance for enhancing acidogenic biohydrogen production. Bioresour Technol 165:372–382

Chandrasekhar K, Lee Y, Lee D (2015) Biohydrogen production: strategies to improve process efficiency through microbial routes. Int J Mol Sci 16:8266–8293

Coates JD, Bhupathiraju VK, McInerney MI, Lovley DR, Achenbach LA (2001) Geobacter hydrogenophilus, geobacter chapellei and geobacter grbiciae, three new, strictly anaerobic, dissimilatory Fe(III)-reducers. Int J Syst Evol Micr 51:581–588

Cummings DE, Caccavo F, Spring S, Rosenzweig F (1999) Ferribacterium limneticum, gen. nov., sp. Nov., an Fe (III) reducing microorganism isolated from mining-impacted freshwater lake sediments. Arch Microbiol 171(3):183–188

Cusick Roland D, Kiely Patrick D, Logan Bruce E (2010) A monetary comparison of energy recovered from microbial fuel cells and microbial electrolysis cells fed winery or domestic wastewaters. Int J Hydrogen Energy 35:8855–8861

El-Chakhtoura J, El-Fadel M, Ananda Rao H, Ghanimeh S, Saikaly PE, Li D (2014) Electricity generation and microbial community structure of air-cathode microbial fuel cells powered with the organic fraction of municipal solid waste and inoculated with different seeds. Biomass Bioenergy 67(2014):24–31

El Mekawy A, Srikanth S, Bajracharya S, Hegab HM, Singh Nigam P, Singh A, Mohan SV, Pant D (2015) Food and agricultural wastes as substrates for bioelectrochemical system (BES): the synchronized recovery of sustainable energy and waste treatment. Food Res Int 73:213–225

European Commission (2013) Future brief: bioelectrochemical systems, wastewater treatment, bioenergy and valuable chemicals delivered by bacteria. Sci Environ Policy 5. http://ec.europa.eu/science-environment-policy

Faaji A (2006) Modern biomass conversion technologies. Mitigation Adapt Strateg Glob Chang. doi: 10.1007/s11027-005-9004-7

Faaij A, Hekkert M, Worrell E, van Wijk A (1998) Optimization of the final waste treatment system in the Netherlands. Resour Conserv Recy 22:47–82

Favoino E, Hoggs D (2008) The potential role of compost in reducing greenhouse gases. Waste Manage Res 26(1):61–69

Giusti L (2009) A review of waste management practices and their impact on human health. Waste Manag 29:2227–2239

Higgins SR, Lopez RJ, Pagaling E, Yan T, Cooney MJ (2013) Towards a hybrid anaerobic digester-microbial fuel cell integrated energy recovery system: an overview of the development of an electrogenic biofilm. Enzyme Microbial Technol 52:344–351

Ieropoulos I, Melhuish C, Greenman J, Horsfield I (2005) EcoBot-II: an artificial agent with a natural metabolism. Adv Robot Syst 2(4):295–300

Ieropoulos IA, Papaharalabos G, Ledezma P, Melhuish C, Stinchcombe A, Greenman J (2013) Waste to real energy: the first MFC powered mobile phone. Phys Chem Chem Phys 15:15312–15316

Jung SP, Yoon MH, Lee SM, Oh SE, Kang H, Yang JK (2014) Power generation and anode bacterial community compositions of sediment fuel cells differing in anode materials and carbon sources. Int J Electrochem Sci 9:315–326

Karluvali A, Köroglu EO, Manav N, Çetinkaya Afs_in Y, Özkaya B (2015) Electricity generation from organic fraction of municipal solid wastes in tubular microbial fuel cell. Separ Purif Technol 156:502–511

Logan BE, Regan JM (2006) Microbial fuel cells: challenges and applications. Environ Sci Technol 5172–5180

Luo S, Sun H, Ping Q, Jin R, He Z (2016) A review of modeling bioelectrochemical systems: engineering and statistical aspects. Energies 9:111

Mohan VS, Chandrasekhar K (2011) Solid phase microbial fuel cell (SMFC) for harnessing bioelectricity from composite food waste fermentation: influence of electrode assembly and buffering capacity. Bioresour Technol 102:7077–7085

Montpart N, Rago L, Baeza JA, Guisasola A (2015) Hydrogen production in single chamber microbial electrolysis cells with different complex substrates. Water Res 68:601–615

Morris JM, Jin S, Crimi B, Pruden A (2009) Microbial fuel cell in enhancing anaerobic biodegradation of diesel. Chem Eng J 146:161–167

Nastro RA, Dumontet S, Ulgiati S, Falcucci G, Vadursi M, Jannelli E, Minutillo M, Cozzolino R, Trifuoggi M, Erme G, De Santis E (2013) Microbial fuel cells fed by solid organic waste: a preliminar experimental study. In: European Fuel Cell Conference EFC 2013 – Rome, 11-13 December. Oral presentation. Article in book of proceedings, pp 139–140

Nastro RA (2014) Microbial fuel cells in waste treatment: recent advances. Int J Perform Eng 10 (4):367–376

Nastro RA, Falcucci G, Toscanesi M, Minutillo M, Pasquale V, Trifuoggi M, Dumontet S, Jannelli E (2015a) Performances and microbiology of a microbial fuel cell (MFC) fed with the organic fraction of municipal solid waste (OFMSW). In: Proceedings of EFC15 Conference, Naples (Italy), 15–18 Dec 2015

Nastro RA, Falcucci G, Hodgson DM, Minutillo M, Trifuoggi M, Guida M, Avignone-Rossa Dumontet S, Jannelli E, Ulgiati S (2015b) Utilization of agro-industrial and urban waste as feedstock in microbial fuel cells (MFCs). In: Proceedings of the global cleaner production and sustainable consumption conference, Sitges (Spain), 1–4 Nov 2015

Oh ST, Kim JR, Premier GC, Lee TH, Kim C, Sloan WT (2010) Sustainable wastewater treatment: how might microbial fuel cells contribute. Biotechnol Adv 28:871–881

Oyetunde T, Ofiteru D, Rodriguez J (2013) Modeling bioelectrochemical systems for wastewater treatment and bioenergy recovery with COMSOL multiphysics®. In: Proceedings of the 2013 COMSOL conference in Boston

Pant D, Van Bogaert G, Diels L, Vanbroekhoven K (2010) A review of the substrates used in microbial fuel cells (MFCs) for sustainable energy production. Bioresource Technol 101:1533–1543

Pant D, Singh A, Van Bogaert G, Alvarez Gallego Y, Diels L, Vanbroekhoven K (2011) An introduction to the life cycle assessment (LCA) of bioelectrochemical systems (BES) for sustainable energy and product generation: relevance and key aspects. Renew Sust Energ Rev 15:1305–1313

Pant D, Singh A, Vanbroekhoven K, Van Bogaert G, Irving Olsen S, Singh Nigam P, Diels L (2012) Bioelectrochemical systems (BES) for sustainable energy production and product recovery from organic wastes and industrial wastewaters. RSC Adv 2:1248–1263

Porta D, Milani S, Lazzarino AI, Perucci CA, Forastiere F (2009) Systematic review of epidemiological studies on health effects associated with management of solid waste. Environ Health 8–60

Premier GC, Kim JR, Massanet-Nicolau J, Kyazze G, Esteves SRR, Penumathsaa BKV, Rodríguez J, Maddy J, Dinsdale RM, Guwy AJ (2012) Integration of biohydrogen, biomethane and bioelectrochemical systems. Renew Energy 49:188–192

Rabaey K, Rozendal RA (2010) Microbial electrosynthesis—revisiting the electrical route for microbial production. Nat Rev Microbiol 8(10):706–716

Recio-Garrido D, Perrier M, Tartakovsky B (2016) Modeling optimization and control of bioelectrochemical systems. Chem Eng J 289:180–190

Rushton L (2003) Health hazards and waste management. Br Med Bull 68:71

Shaoan LH, Chen Chen G (2011) Bioelectrochemical systems for efficient recalcitrant wastes treatment. J Chem Technol Biotechnol 86:481–491

Sleutels THJA, Ter Heijne A, Buisman CJN, Hamelers HVM (2012) Bioelectrochemical systems: an outlook for practical applications. ChemSusChem 5:1012–1019

UNEP - United Nations Environmental Programme (2010) Waste and Climate Change - Global trends and strategy framework. Available on http://www.unep.or.jp/ietc/Publications/spc/Waste&ClimateChange/Waste&ClimateChange.pdf. Accessed Dec 2015

Van Eerten-Jansen MCAA, Ter Heijne A, Buisman CJM, Hamelers HVM (2012) Microbial electrolysis cells for production of methane from CO2: long-term performance and perspectives. Int J Energy Res 36:809–819

Villano M, Monaco G, Aulenta F, Majone M (2011) Electrochemically assisted methane production in a biofilm reactor. J Power Sources 196:9467–9472

Wang C, Lee Y, Liao F (2015) Effect of composting parameters on the power performance of solid microbial fuel cells. Sustainability 7(9):12634–12643

Wei J, Liang P, Huang X (2011) Recent progress in electrodes for microbial fuel cells. Bioresour Technol 102:9335–9344

Winfield J, Chambers LD, Rossiter J, Stinchcombe A, Walter XA, Greenman J, Ieropoulos I (2015) Fade to green: a biodegradable stack of microbial fuel cells. ChemSusChem 8:2705–2712

Zeitfracht Medien GmbH
Ferdinand-Jühlke-Straße 7
99095 Erfurt, Deutschland
produktsicherheit@kolibri360.de